# Getting Started with BizTalk Services

Create powerful integration solutions for the cloud using
the extensible Windows Azure BizTalk Services

**Karthik Bharathy**

**Jon Fancey**

[PACKT] enterprise

PUBLISHING    professional expertise distilled

BIRMINGHAM - MUMBAI

# Getting Started with BizTalk Services

First published: March 2014

Production Reference: 1200314

Published by Packt Publishing Ltd.
Livery Place
35 Livery Street
Birmingham B3 2PB, UK.

ISBN 978-1-78217-740-1

www.packtpub.com

Cover Image by Jarek Blaminsky (milak6@wp.pl)

# Credits

**Authors**
Karthik Bharathy

Jon Fancey

**Reviewers**
Steef-Jan Wiggers

Kevin Smith

Tomas Restrepo

Daniel Bullington

**Acquisition Editor**
Joanne Fitzpatrick

**Content Development Editor**
Shaon Basu

**Technical Editors**
Kunal Anil Gaikwad

Pramod Kumavat

Venu Manthena

Mukul Pawar

Siddhi Rane

**Copy Editors**
Janbal Dharmaraj

Sayanee Mukherjee

Karuna Narayanan

Adithi Shetty

**Project Coordinator**
Aboli Ambardekar

**Proofreader**
Simran Bhogal

Paul Hindle

**Indexer**
Priya Subramani

**Graphics**
Ronak Dhruv

Abhinash Sahu

**Production Coordinator**
Pooja Chiplunkar

**Cover Work**
Pooja Chiplunkar

# Foreword

The cloud moves fast.

Welcome to Windows Azure BizTalk Services, a key part of Microsoft's cloud integration vision. With BizTalk Services, customers can connect their businesses through process automation and integrate the Web with their existing backend systems on premises; all from a flexible and scalable rock-solid platform managed using industry-leading tools.

With integration, it's not a cloud or on-premises decision, it's both. This book provides you with a great introduction to BizTalk Services, a fantastic new cloud service from Microsoft designed to help you get started quickly and productively in the shortest time possible. Jon and Karthik have done a great job in making the material easy to grasp for newcomers to the Microsoft stack as well as those experienced in BizTalk Server looking to start using BizTalk Services.

I really liked Karthik's and Jon's writing style and found this book an excellent introduction to BizTalk Services. Unlike other books, this doesn't attempt to be a huge reference full of extraneous details. Instead, it offers a relatively quick and concise read that details how to use the most important features. The result is a very approachable book that provides a great way to learn BizTalk Services and how to immediately take advantage of it.

Hold on and enjoy the ride!

**Scott Guthrie**

Corporate Vice President, Windows Azure, Microsoft Corporation

March 2014

# Foreword

Karthik and I have worked together for over two years on Windows Azure BizTalk Services. It is a pleasure for me to introduce this book on Windows Azure BizTalk Services on behalf of Karthik and Jon.

BizTalk has been a leader in the integration space for many years now and is used by a majority of the world's largest companies for their mission-critical-systems integration. As cloud computing changes how enterprises run their business, it is important to bring this product to the cloud as a part of Windows Azure Platform. We started building Windows Azure BizTalk Services for cloud-to-cloud and cloud-to-on-premises integration scenarios, targeting both enterprise and SMB customers. This new "built from ground up for cloud" service will harness all the power of cloud and yet make it simple to use.

In the world of modern applications and services, there is a need for IT admins and developers to comprehensively understand a technology such as BizTalk Services and apply it in their own IT ecosystem. This book precisely fills this need. This book showcases many practical, real-world scenarios and provides detailed hands-on walkthroughs of cloud integration to allow the reader to quickly understand the material presented.

I invite you to join the integration journey with Karthik and Jon as they uncover the capabilities of BizTalk Services in a lucid, approachable manner. I hope you will enjoy the book as much as I did and that it helps you to leverage BizTalk Services more effectively in your organization.

**Vivek Dalvi**

Principal Group Program Manager, BizTalk Product Group

March 2014

# About the Authors

**Karthik Bharathy** is a Lead Program Manager in the BizTalk product group with nearly a decade of software experience. He has been with the product group since the days when BizTalk Services started off as a set of ideas on the drawing board. He has also shipped releases of BizTalk Server, SQL Server, and Visual Studio. In his current role, he oversees the B2B platform experience across industry verticals. He has presented at several Microsoft conferences, including BizTalk Summit US and Europe, TechEd EMEA and US, TechReady US, MVP Summit US, and TechDays India.

His passion for computers started at the age of 12 when he coded BASIC on the ZX Spectrum. He graduated from Bangalore University in Computer Science summa cum laude and also holds a management degree from the Indian School of Business. In his spare time, he is usually travelling and is a major foodie.

To all the members of the BizTalk family—the product group, CSS, DPE, CAT, Marketing, UE, and UX—thank you for building this awesome integration product called BizTalk. I truly appreciate the complexity of the middleware breathing BizTalk every day! I would like to thank Vivek Dalvi, Sandeep Prabhu, Shridhar Diwan, and Rajesh Ramamirtham for the discussions and support while writing this book.

I would like to thank the MVP community for their discussions on BizTalk. I learned a lot from you guys and I respect the level of commitment you inspire towards the product. I would like to thank Steef-Jan Wiggers, Richard Seroter, Michael Stephenson, Sarvana Kumar, Sandro Pereira, Kent Weare, Mick Badran, Rick Garibay, Stephen Thomas, Bill Chestnut, Sam Vanhoutte, Dwight Goins, Ben Cline, and Mikael Hakansson for their constant feedback on BizTalk.

Special thanks to Scott Guthrie for agreeing to write the foreword of this book, and thanks to our Content Development Editor Shaon Basu and Project Coordinator Aboli Ambardekar, whose reminders and feedback kept us on our toes and helped us land the book on time.

Above all, I would like to thank my wife, Thulasi and my parents who supported and encouraged me throughout this journey.

**Jon Fancey** is an integration veteran who has worked on the Microsoft stack for over 20 years. He is a nine-time Microsoft Integration MVP and has worked closely with both the BizTalk Server and Host Integration Server product groups for nearly a decade. He has presented at many major conferences including TechEd, DevWeek, and the 2014 London BizTalk Summit. He has also written numerous articles and whitepapers for MSDN on BizTalk, SharePoint, and other topics.

Jon co-founded Affinus, a UK-based Microsoft partner, with Kevin B. Smith, formerly from the BizTalk product group, shipping the first three versions of the BizTalk product. Affinus works closely with large enterprise customers on interesting integration challenges, helping them move to the cloud.

Jon lives in West Sussex, UK, with his wife Fiona and two children, Ben and Tom, and their dog, Dilly.

I would like to thank the following for their help and assistance: Kevin Smith, Tomas Restrepo, and Steef-Jan Wiggers for tirelessly reviewing every word of this book, especially Steef-Jan who turned it round in a weekend on very tight deadlines. You all made it a better book, and I am very grateful for that. I'd also like to thank other members of the Affinus family, Daniel Probert and Simon Poulter, who've put up with me discussing this project for a very long time and guided my thinking along the way.

A special thanks goes to Scott and Vivek for agreeing to write forewords for us, your support on this project has been very much appreciated. And of course thanks to everyone at Packt for their support and encouragement, and deadline management!

Finally, I'd like to thank my family for giving me the time and space to write my first book; I know it's tough sometimes and without your encouragement and support this project wouldn't have been possible.

# About the Reviewers

**Steef-Jan Wiggers** has over 15 years of experience as a technical lead developer, application architect, and consultant, specializing in custom applications, enterprise application integration (BizTalk), web services, and Windows Azure. He is very active in the BizTalk community (`http://social.technet.microsoft.com/wiki/contents/articles/7141.user-page-steef-jan-wiggers-microsoft-biztalk-server-consultant-and-mvp.aspx`) as a blogger, Wiki author/editor, forums writer, and public speaker in the Netherlands and Europe. For these efforts, Microsoft has recognized him as a Microsoft MVP for the past four years. On his personal blog (`http://soa-thoughts.blogspot.com/`) and BizTalk Administrators blog (`http://www.biztalkadminsblogging.com/`), he shares his knowledge about SOA, Azure (Service Bus), BizTalk Services, and BizTalk.

In addition to consulting, he is also an author and has been a technical reviewer for Packt Publishing. He has written the book *BizTalk Server 2010 Cookbook*, *Packt Publishing*, and has technically reviewed the following books:

- *Microsoft BizTalk Server 2010 Patterns* by Dan Rosanova
- *(MCTS): Microsoft BizTalk Server 2010 (70-595) Certification Guide* by Johan Hedberg, Morten la Cour, and Kent Weare

> Windows Azure BizTalk Services is a new service in Azure and a promising technology for integration (EAI) and B2B in the cloud. This book provides readers with background information and hands-on experience working with BizTalk Services. I would like to thank the authors Jon Fancey and Karthik Bharathy, both of whom I know personally, for giving me the opportunity to review this book. They have done an excellent job writing it.

**Kevin Smith** is a co-founder of Affinus, a UK-based Microsoft partner and previously worked in the BizTalk Server product group for six years delivering BizTalk Server 2000, 2002, and the seminal third release 2004, which created the much-praised BizTalk architecture that the current product is based on. Kevin works on hard .NET problems for customers and specializes in the investment banking industry. His primary interests lie in UX design and machine learning.

**Tomas Restrepo** has been writing software for over 10 years, starting with C/C++ and eventually moving to the .NET platform. He currently spends most of his time helping other developers solve complex problems and troubleshooting application performance and scalability issues.

**Daniel Bullington** is a technology architect, manager, and strategist with industrial experience in financial services, healthcare, management consulting, and Software as a Service (SaaS), working for several well-known Fortune 500 and Fortune 50 companies. His focus has been on large-scale web/mobile, SOA/EAI, DW/BI, and cloud solutions. Daniel drives continuous improvement and operational excellence (including an intelligent level of process, metrics/KPIs, and so on) to spur meaningful IT organizational change and an enhanced value proposition through positive business outcomes.

# www.PacktPub.com

## Support files, eBooks, discount offers and more

You might want to visit www.PacktPub.com for support files and downloads related to your book.

Did you know that Packt offers eBook versions of every book published, with PDF and ePub files available? You can upgrade to the eBook version at www.PacktPub.com and as a print book customer, you are entitled to a discount on the eBook copy. Get in touch with us at service@packtpub.com for more details.

At www.PacktPub.com, you can also read a collection of free technical articles, sign up for a range of free newsletters and receive exclusive discounts and offers on Packt books and eBooks.

http://PacktLib.PacktPub.com

Do you need instant solutions to your IT questions? PacktLib is Packt's online digital book library. Here, you can access, read and search across Packt's entire library of books.

## Why Subscribe?

- Fully searchable across every book published by Packt
- Copy and paste, print and bookmark content
- On demand and accessible via web browser

## Free Access for Packt account holders

If you have an account with Packt at www.PacktPub.com, you can use this to access PacktLib today and view nine entirely free books. Simply use your login credentials for immediate access.

## Instant Updates on New Packt Books

Get notified! Find out when new books are published by following @PacktEnterprise on Twitter, or the *Packt Enterprise* Facebook page.

# Table of Contents

# Preface

It all started about a year ago and BizTalk Services was soon to go for preview in a few months. We were all excited to break new ground in the era of cloud middleware. We must tell you one of the benefits of being in the product group (or being an MVP) is that you get early access to bits long before they hit the market. Working on those bits, we thought to ourselves, "Wouldn't it be nice for our customers to have a guide to build effective solutions with this service?" This book on BizTalk Services was envisioned not necessarily to spoil the fun by adding every little detail, but to cover enough to understand the architecture, the key components, and help you explore.

This book is written for beginners, and knowledge of BizTalk Server is neither assumed nor expected. It is also the first book on the topic, and we'll cover all the important features including EAI, B2B, and hybrid deployments in detail—all with code samples and walkthroughs. If you are an EAI user, you can start with *Chapter 1, Hello BizTalk Services* and then continue with *Chapter 2, Messages and Transforms*, *Chapter 3, Bridges*, and *Chapter 4, Enterprise Application Integration*. On the other hand, a B2B developer or architect can follow *Chapter 1, Hello BizTalk Services*, *Chapter 2, Messages and Transforms*, and *Chapter 5, Business-to-business Integration*. If you're interested in the APIs underpinning the services, troubleshooting your solutions, or how to move to BizTalk Services, then *Chapter 6, API*, *Chapter 7, Tracking and Troubleshooting*, and *Chapter 8, Moving to BizTalk Services* will guide you.

# What this book covers

*Chapter 1, Hello BizTalk Services,* introduces BizTalk Services, its architecture, and how to create an instance of the service and deploy solutions.

*Chapter 2, Messages and Transforms,* explains message processing and how to transform messages to different formats. Also, it explains how to use mapping operations to aggregate data, perform reference data lookups, and use custom code in transformations.

*Chapter 3, Bridges,* gives a detailed look at bridges and explains how to enrich messages and route messages to different endpoints.

*Chapter 4, Enterprise Application Integration,* explains sources and destinations and how to connect BizTalk Services to enterprise applications and systems on-premises from the cloud.

*Chapter 5, Business-to-business Integration,* discusses B2B integration using industry standard protocols such as EDIFACT, X12, and AS2. It also discusses how to create partners and agreements in BizTalk Services to connect with trading partners and how to utilize message batching and archiving.

*Chapter 6, API,* discusses a rich API underpinning BizTalk Services. Also, it explains what it can do and how to use it in different contexts, including REST, PowerShell, and custom code.

*Chapter 7, Tracking and Troubleshooting,* discusses how messages are tracked in BizTalk Services and how to find and resolve problems when they occur using the tools BizTalk Services provides.

*Chapter 8, Moving to BizTalk Services,* explains how to move from BizTalk Server to BizTalk Services, the differences between the two products, and future plans.

# What you need for this book

To follow along with the code samples and solutions provided in the book, you will need the following pre requisites:

- Internet access
- One of the following operating systems: Windows 7 Service Pack 1, Windows 8, Windows 8.1, Windows Server 2008 R2 SP1, Windows Server 2012, or Windows Server 2012 R2
- Internet Explorer 9 or Internet Explorer 10

- Visual Studio 2012
- Windows Azure BizTalk Services SDK

  Please visit `http://msdn.microsoft.com/en-us/library/windowsazure/hh689760.aspx`

- A Windows Azure subscription and instance of Windows Azure BizTalk Services

  To create a BizTalk Service instance, please visit `http://www.windowsazure.com/en-us/pricing/free-trial/`

## Who this book is for

This book is for software developers, IT pros, architects, and technical managers who wish to understand BizTalk Services, what it can do, and how to use it to integrate services, on-premises applications, and businesses together.

## Conventions

In this book, you will find a number of styles of text that distinguish between different kinds of information. Here are some examples of these styles, and an explanation of their meaning.

Code words in text, database table names, folder names, filenames, file extensions, pathnames, dummy URLs, user input, and Twitter handles are shown as follows: "Let's tackle the `ShippingAddress` node."

A block of code is set as follows:

```
public string CreateAddress (string Number, string Street, string
City, string State, string Country)
{
  return Number + " " +
    Street + "," +
    City + "," +
    State + "," +
    Country;
}
```

Any command-line input or output is written as follows:

```
select-azuresubscription –SubscriptionName "Test"
```

**New terms** and **important words** are shown in bold. Words that you see on the screen, in menus, or dialog boxes for example, appear in the text like this: "Click on **Add** to create the map and add it to the solution."

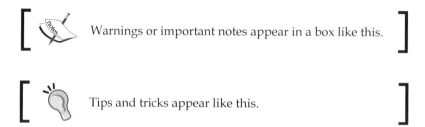

> Warnings or important notes appear in a box like this.

> Tips and tricks appear like this.

# Reader feedback

Feedback from our readers is always welcome. Let us know what you think about this book—what you liked or may have disliked. Reader feedback is important for us to develop titles that you really get the most out of.

To send us general feedback, simply send an e-mail to feedback@packtpub.com, and mention the book title via the subject of your message.

If there is a topic that you have expertise in and you are interested in either writing or contributing to a book, see our author guide on www.packtpub.com/authors.

# Customer support

Now that you are the proud owner of a Packt book, we have a number of things to help you to get the most from your purchase.

## Downloading the example code

You can download the example code files for all Packt books you have purchased from your account at http://www.packtpub.com. If you purchased this book elsewhere, you can visit http://www.packtpub.com/support and register to have the files e-mailed directly to you.

## Errata

Although we have taken every care to ensure the accuracy of our content, mistakes do happen. If you find a mistake in one of our books—maybe a mistake in the text or the code—we would be grateful if you would report this to us. By doing so, you can save other readers from frustration and help us improve subsequent versions of this book. If you find any errata, please report them by visiting http://www.packtpub.com/submit-errata, selecting your book, clicking on the **errata submission form** link, and entering the details of your errata. Once your errata are verified, your submission will be accepted and the errata will be uploaded on our website, or added to any list of existing errata, under the Errata section of that title. Any existing errata can be viewed by selecting your title from http://www.packtpub.com/support.

## Piracy

Piracy of copyright material on the Internet is an ongoing problem across all media. At Packt, we take the protection of our copyright and licenses very seriously. If you come across any illegal copies of our works, in any form, on the Internet, please provide us with the location address or website name immediately so that we can pursue a remedy.

Please contact us at copyright@packtpub.com with a link to the suspected pirated material.

We appreciate your help in protecting our authors, and our ability to bring you valuable content.

## Questions

You can contact us at questions@packtpub.com if you are having a problem with any aspect of the book, and we will do our best to address it.

# Hello BizTalk Services

As companies build and extend their IT information assets, there is a need to integrate applications to exchange data. While a point-to-point application integration using custom code is possible, it does not address large-scale application integrations without duplication of data and the complexity of managing such an integration. BizTalk is a de facto choice for message-based integration for on-premises systems and for services on Azure for both applications and businesses.

In this book, we're going to introduce you to BizTalk Services, Microsoft's new integration middleware hosted on Windows Azure. This book assumes that you already understand the need for integration and the many benefits of using specialist integration software as opposed to building custom point-to-point integration solutions. We will look at why cloud-hosted integration services are so compelling and the scenarios in which these services make most sense. Prior knowledge of BizTalk Server is not expected, though it will help you to quickly understand certain topics within the book. We will assume you are a professional developer, an architect, or a solution designer familiar with the challenges of integrating heterogeneous systems and applications. The book also assumes familiarity of the Microsoft developer toolset, specifically .NET, Visual Studio, and SQL.

In this chapter, we will start to answer some fundamental questions around what BizTalk Services is and how it can help you to build integration solutions. Specifically, the following topics will be discussed:

- The business and technical drivers for BizTalk Services on Azure
- Concepts and architecture of BizTalk Services
- Using BizTalk Services to realize a simple purchase order scenario

# Background

First though, some background is necessary. At the **Professional Developers Conference (PDC)** 2008, Microsoft unveiled Windows Azure—a new operating system designed for the cloud. Over the subsequent years, Windows Azure has become Microsoft's de facto cloud platform covering services, media, websites, mobile applications, and more. While BizTalk Server has been an established product for over ten years (and eight releases), cloud adoption has been driving connected systems across different services and Line-of-Business applications. To meet the growing demand for cloud-based integration, the Microsoft BizTalk team released the first version of BizTalk Services, named Service Bus EAI and EDI Labs Community Technology Preview (CTP) on December 17, 2011. The goal was for customers to be able to sign up in a shared environment and set up simple XML/EDI flows without worrying about installation and maintenance. The capabilities were rich enough to enable simple point-to-point integration scenarios; on April 9th of the following year, the CTP was refreshed, incorporating feedback from customers by providing additional capabilities.

The CTP environment was hosted on a publicly shared service, and therefore had restrictions on running users' custom code. Integration is rarely straightforward, and the ability for developers to write custom code and deploy it as part of their integration solutions was a key requirement. Customers had also needed guarantees around performance and Service Level Agreements (SLAs). Hosting user assemblies as part of the cloud services became a reality by switching to per tenant deployment. Thus, the cloud offering known as BizTalk Services was born on June 3, 2013. Like many Azure services, BizTalk Services is expected to be updated at a regular cadence. The latest update as of this writing was on February 20, 2014. This technology is opening up new integration possibilities; with Microsoft's on-going investments, it will be on par with the capabilities of its on-premises cousin, BizTalk Server, in the near future.

# Business drivers

There are many tangible benefits of building solutions on Azure today. A few of these include the ability to scale up/down the platform based on predictable or dynamic shifts in application load and throughput requirements without worrying about the hardware procurement time and setup, the ability to pay as you go with expense incurred as an Operational Expenditure (OpEx) instead of a Capital Expenditure (CapEx), and the increase in reach for certain integration scenarios.

Specific to integration on Azure, there are four factors that drive adoption:

- **Focus on business operations, not IT**: There are business benefits in terms of reduced cost by leveraging platforms running under economy of scale, making it cheaper for customers to obtain them with higher quality of services. The need of the hour is to simplify IT deployment and management and focus on business services instead of configuring software or hardware.

- **Simplicity of managing externally facing services**: Enterprises typically offer services across geographic and organizational boundaries. Many of these services require making changes to the corporate firewall to allow or deny the applications. This process is an IT nightmare for many organizations. With Azure integration services such as B2B, which inherently require external communication, all this could be moved to the cloud and the necessary access and control policies could then be centralized for all internal services. This also enables self-service configuration changes to the application, thereby dramatically improving responsiveness to business changes.

- **Greenfield cloud applications**: The proliferation of mobile devices, such as smartphones and tablets as well as Point-of-Sale (POS) systems, has given rise to services that are inherently cloud based. Think of a POS that transmits daily transaction logs to its backend Line-of-Business systems or an RFID service that transmits information about each item in a shopping cart purchased in a retail store to an inventory application. As new services are being rolled out, organizations want to be able to develop and deploy these services with shorter time-to-market using Azure.

- Another factor to consider for the cloud is to leverage existing on-premises investments. Businesses have invested in a variety of on-premises systems, including traditional ERP as well as legacy mainframes and other bespoke systems that literally run the business.

# Technical drivers

The primary technical goal for BizTalk Services is to reduce the impedance mismatch between source and destination systems that are exchanging information. Such impedance can be at different levels:

- **Transport protocol impedance**: The source might send messages over one transport (say FTP) and the destination may only accept messages over another transport (say POP3). It could also be the case that messages are sent from one LOB to another, for example, one end is sending messages from the Sales Force adapter while the other end is accepting messages via the SAP adapter. BizTalk Services provides the notion of adapters to resolve this impedance.

- **Application protocol impedance**: The source might send EDIFACT messages while the destination may only accept messages in XML. BizTalk Services provides native support to protocols such as X12, EDIFACT, and flat files to resolve this impedance.

- **Format impedance**: The source might send messages in one XML format while the destination may only accept messages in another XML format. BizTalk Services provides transforms to resolve this impedance.

- **Timing impedance**: The source can send messages any time of the day, but the destination only accepts messages between 4 and 7 P.M. The source can send messages twice as fast as the destination can process them.

- **Size impedance**: The source can send messages of any size, but the destination can accept messages of 1 MB at most.

BizTalk Services provides connectivity to Service Bus, batching and debatching to resolve the last two impedances.

# Core scenarios

The aforementioned drivers have resulted in three core scenarios for BizTalk Services:

- **Enterprise Application Integration**: These are primarily messaging scenarios with flat file or XML-based data that are between two or more applications, atleast one of which is running in the cloud. A good example would be a travel portal connected to the ticketing systems of multiple airlines.

- **Business-to-Business Integration**: These are messaging scenarios with structured flat file/XML between two organizations. An example would be an IT company procuring hardware from vendors such as HP, Dell, or Lenovo.

- **Connectivity with Hybrid Applications**: These are messaging scenarios between Azure and on-premises applications. An example here would be connecting a Salesforce application to a SAP application running in your internal IT environment.

We will look at these scenarios in detail in separate chapters of this book.

# Concepts

The following figure illustrates a basic integration flow from an FTP source to a LOB destination. BizTalk Services, represented by the middle box, is a sequence of processing steps.

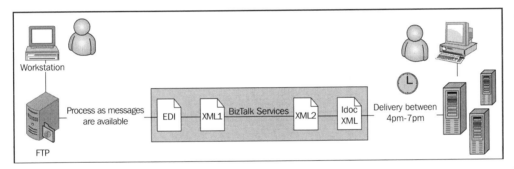

Role of BizTalk Services

BizTalk Services introduces several key concepts to facilitate building integration solutions on Azure:

- **Bridge**: A bridge is a unit of processing in BizTalk Services that can address impedance mismatch. It contains three units: one or more source locations (for example, FTP) to read messages from, a pipeline to process the message, and one or more destinations (for example, Queue) to write the processed messages. The pipeline is divided into distinct processing units called stages, each with its own function (for example, a stage in a pipeline can validate a message against a schema). A series of stages represents the bridge pattern or bridge template. Out of the-box, BizTalk Services v1 ships with three templates: XML, EDI, and AS2.

- **Adapter**: An adapter is the transport medium that can send messages (to a destination) or receive messages (from a source) and pass them to the pipeline in a bridge; for example, Line-of-Business adapters such as SAP and Oracle EBS or transport adapters such as FTP and SFTP.

- **Transform**: A transform converts a message from one format into another, aiding structural conversion. Transforms contain operations that can perform commonly used transformations like string operations, loop constructs, list operations, and arithmetic and logical expressions.

- **Application protocol**: A protocol defines the message format and processing semantics such as the requirement to send and correlate acknowledgements of messages.

- **Route**: A route defines the destination endpoint where the message will be sent based on the specified criteria. The route criteria are evaluated based on SQL-92 expression syntax.

- **Batching**: The aggregation of messages based on selection criteria is termed batching. The release (sending) of a batch is governed by size, count, or time, or a combination of these parameters.

- **Promoted properties**: Promoted properties are name-value pairs, where the name is user-defined and the value is derived from the message header, message body, or from the context within the bridge. Promoted properties are commonly used in batching and routing to specify their criteria.

- **Artifact**: An artifact is anything that aids in the processing of the message in the bridge. XML schemas, maps, custom assemblies, and certificates are the artifacts used in XML and EDI bridges. In BizTalk Services, each artifact is stored in the artifact store and is addressable by a unique URL.

# Life cycle and architecture

Unlike most other Azure services deployments, BizTalk Services provisions dedicated resources for storage and compute instances that are isolated across tenants. This means that no two deployments have anything in common between them. The advantage is that you can write any custom code and be assured that you cannot impact the performance or availability of other deployments. This dedicated deployment also offers isolation of data at the storage level, thus increasing the privacy of data and SLA of the service.

Broadly categorizing, there are three steps to go through before you can have an active usable deployment. The first is to provision the service using standard Azure tools, including the Azure portal to create the service; the second is to deploy the necessary artifacts and configuration outlined in the *BizTalk Services concepts* section using Visual Studio and the BizTalk Services portal; and the third is sending or receiving the messages.

The architecture of BizTalk Services contains three key components:

- **Provisioning services**: This is a set of Microsoft services that manage the lifecycle of a BizTalk Services deployment as well as monitor its health. It also includes components to bill the end user based on usage of BizTalk Services. The management interface to the service is exposed via the **Red Dog Front End (RDFE)** public API. The Azure Management Portal or PowerShell scripts from the user go via the RDFE API. Using the service, you can scale up/down your deployment as well as back up and restore deployment across datacenters.

 Red Dog was the original codename for Azure, with the "FE" being the publicly accessible frontend that users directly interact with either via the Azure portal or service management APIs.

- **Per tenant BizTalk Services**: This is the per tenant deployment that is created in the user's Azure subscription. A BizTalk Services deployment is identified by the deployment name and is accessible using the URL secured by the Access Control Service (ACS). All artifacts such as bridges and schemas are added into the deployment with a URL which is a sub-path of the deployment URL. For example, a bridge is added under `<deployment URL>/default/<bridgeName>`. Here default is the namespace name where the artifacts are grouped into.

- **Per tenant dependencies**: These dependencies are the Azure services required for tracking, troubleshooting, and security. For example, BizTalk Services provides a tracking store, which is an Azure SQL database where the processing status of messages, along with related properties, are stored as the messages pass through a bridge. The information from the tracking store is shown in the BizTalk Services portal tracking view. Archiving and monitoring is stored in Azure storage blobs and tables. Archived messages are stored in blob containers based on the date of the archive. The storage also contains the Azure table WADLogsTable, where tracing information for a bridge can be obtained. Finally, Access Control Service regulates access to all endpoints in the deployment. During deployment creation, the provisioning service uses the Management Service credentials to programmatically access ACS to create a relying party for the BizTalk Services deployment, add rule groups for Send, Listen, and Manage claims, and create the service identity with the necessary passwords for directly talking to ACS for deployment of the artifacts. The interaction between these components is illustrated in the following figure:

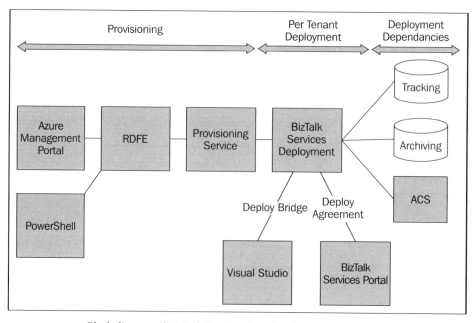

Block diagram of BizTalk Services shared and per tenant services

# Personas and tools

BizTalk Services provides different user experiences for different personas to facilitate an optimized, task-based approach. The principal personas are:

| Persona | Description | Primary Tools |
|---|---|---|
| Developer | Someone who creates integration solutions and artifacts, such as transforms and schemas | Visual Studio |
| IT Pro | Someone who manages the environment, including tasks such as deployment, setup, and configuration | Azure Portal and dashboard |
| Partner Administrator | Someone who sets up and manages trading partners | Trading partner/ BizTalk portal |

# Developer

A developer will typically focus on creating solutions using Visual Studio. BizTalk Services VS 2012 project templates are provided to enable rapid creation of both EAI and EDI solutions. These provide a graphical work surface on which to create and configure bridges to facilitate communication between an enterprise and its trading partners. Additionally, sophisticated tools are provided, including a graphical mapping interface and schema editors.

# IT Pro

The Windows Azure Platform Management Portal provides access to create the BizTalk Services deployment and management tasks as well as at-a-glance status information providing details on the overall health of all deployments and accounts. In addition to services deployment, the Windows Azure Platform Management Portal provides an interface for provisioning Azure SQL databases, mobile services, Service Bus entities, and so on.

# Partner Administrator

The Partner Administrator persona uses the BizTalk Services portal for a number of management functions such as the creation and administration of trading partners, configuration of agreements including required transformations, routing and acknowledgements, tracking of messages, and exception processing.

The BizTalk Services portal enables the creation of trading partners and agreements between them. This enables the setting up and management of the protocols used to exchange data (for example, X12 and AS2) and the message formats to use together with transformation and routing capabilities. In this way, trading partners can be onboarded and configured quickly and easily by non-IT personnel without the use of developer tools such as Visual Studio, all through the web-based portal.

In addition, the management portal also provides the ability to set and view tracking data on message flows, including both contextual details (sender, message type, and so on) as well as the message bodies themselves. The ability to archive and export message data is also provided as part of the service.

Additionally, RESTful APIs are implemented to provide full fidelity with the portal, enabling activities to be scripted and deployment to be automated. Additionally, integration with customer systems and tools such as SharePoint for tracking data, visualization, or on-premises storage is also possible using this API.

# Deployment considerations

You will need to consider the BizTalk Services edition required for your production use as well as the environment for test and/or staging purposes. This depends on decision points such as:

- Expected message load on the target system
- Capabilities that are required now versus 6 months down the line
- IT requirements around compliance, security, and DR

The list of capabilities across different editions is outlined in the Windows Azure documentation page at `http://www.windowsazure.com/en-us/documentation/articles/biztalk-editions-feature-chart`.

**Note on BizTalk Services editions and signup**

BizTalk Services is currently available in four editions: Developer, Basic, Standard, and Premium, each with varying capabilities and prices. You can sign up for BizTalk Services from the Azure portal. The Developer SKU contains all features needed to try and evaluate without worrying about production readiness. We use the Developer edition for all examples in this book.

# Provisioning BizTalk Services

BizTalk Services deployment can be created using the Windows Azure Management Portal or using PowerShell. We will use the former in this example.

# Certificates and ACS

Certificates are required for communication using SSL, and Access Control Service is used to secure the endpoints of the BizTalk Services deployment. First, you need to know whether you need a custom domain for the BizTalk Services deployment. In the case of test or developer deployments, the answer is mostly no. A BizTalk Services deployment will autogenerate a self-signed certificate with an expiry of close to 5 years. The ACS required for deployment will also be autocreated. Certificate and Access Control Service details are required for sending messages to bridges and agreements and can be retrieved from the Dashboard page post deployment.

# Storage requirements

You need to create an Azure SQL database for tracking data. It is recommended to use the Business edition with the appropriate size; for test purposes, you can start with the 1 GB Web edition. You also need to pass the storage account credentials to archive message data. It is recommended that you create a new Azure SQL database and Storage account for use with BizTalk Services only.

# The BizTalk Services create wizard

Now that we have the security and storage details figured out, let us create a BizTalk Services deployment from the Azure Management Portal:

1. From the Management portal, navigate to **New | App Services | BizTalk Service | Custom Create**.

2. Enter a unique name for the deployment, keeping the following values—**EDITION: Developer, REGION: East US, TRACKING DATABASE: Create a new SQL Database instance**.

3. In the next page, retain the default database name, choose the SQL server, and enter the server login name and password.

>  As of writing this book, there can be six SQL server instances per Azure subscription.

4. In the next page, choose the storage account for archiving and monitoring information.

5. Deploy the solution.

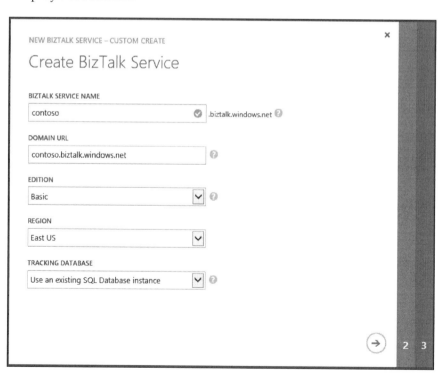

The BizTalk Services create wizard from Windows Azure Management Portal

The deployment takes roughly 30 minutes to complete. After completion, you will see the status of the deployment as Active. Navigate to the deployment dashboard page; click on **CONNECTION INFORMATION** and note down the ACS credentials and download the deployment SSL certificate. The SSL certificate needs to be installed on the client machine where the Visual Studio SDK will be used.

# BizTalk portal registration

We have one step remaining, and that is to configure the BizTalk Services Management portal to view agreements, bridges, and their tracking data. For this, perform the following steps:

1. Click on **Manage** from the Dashboard screen.

2. This will launch `<mydeployment>.portal.biztalk.windows.net`, where the BizTalk Portal is hosted.

3. Some of the fields, such as the user's live ID and deployment name, will be auto-populated.

4. Enter the **ACS Issuer name** and **ACS Issuer secret** noted in the previous step and click on **Register**.

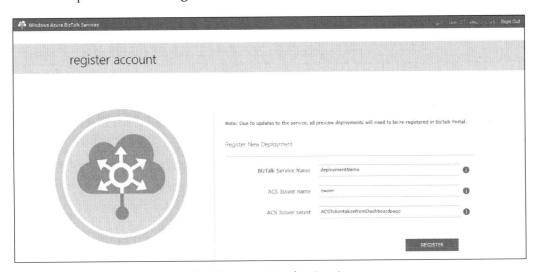

BizTalk Services Portal registration

# Creating your first BizTalk Services solution

Let us put things into action and use the deployment created earlier to address a real-world multichannel sales scenario.

## Scenario description

A trader, Northwind, manages an e-commerce website for online customer purchases. They also receive bulk orders from event firms and corporates for their goods. Northwind needs to develop a solution to validate an order and route the request to the right inventory location for delivery of the goods. The incoming request is an XML file with the order details. The request from event firms and corporates is over FTP, while e-commerce website requests are over HTTP. Post processing of the order, if the customer location is inside the US, then the request are forwarded to a relay service at a US address. For all other locations, the order needs to go to the central site and is sent to a Service Bus Queue at IntlAddress with the location as a promoted property.

## Prerequisites

Before we start, we need to set up the client machine to connect to the deployment created earlier by performing the following steps:

1. Install the certificate downloaded from the deployment on your client box to the trusted root store. This authenticates any SSL traffic that is between your client and the integration solution on Azure.

2. Download and install the BizTalk Services SDK (`https://go.microsoft.com/fwLink/?LinkID=313230`) so the developer project experience lights up in Visual Studio 2012.

3. Download the BizTalk Services EAI tools' Message Sender and Message Receiver samples from the MSDN Code Gallery available at `http://code.msdn.microsoft.com/windowsazure`.

# Realizing the solution

We will break down the implementation details into defining the incoming format and creating the bridge, including transports to process incoming messages and the creation of the target endpoints, relay, and Service Bus Queue.

## Creating a BizTalk Services project

You can create a new BizTalk Services project in Visual Studio 2012.

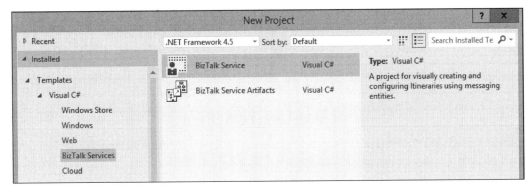

BizTalk Services project in Visual Studio

## Creating the Order schema

From within your project, right-click on the project name, click on **Add** | **New Item**, and add a new **Flat File Schema**.

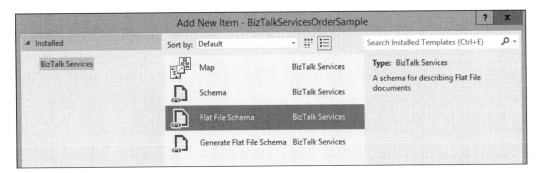

Add Flat File Schema into BizTalk Services project

Add the following nodes to the schema so that the structure looks as follows:

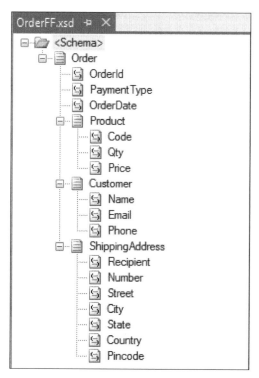

Flat File Schema structure

For each of the records in the XSD file, make sure that the delimiters are added correctly:

```
<b:recordInfo structure="delimited" child_delimiter_type="char" child_
delimiter="," child_order="postfix" preserve_delimiter_for_empty_
data="true" suppress_trailing_delimiters="false" sequence_number="4"
/>
```

You can validate the schema by running it with an instance file. The **Validate Instance** command is available by right-clicking on the created schema file in the Solution Explorer. Add the following flat file and XML instances in two separate files, use the **Validate Instance** command, and verify that the schema validates those instances. For each command run, ensure that the schema properties window has the right **Validate Instance Input Type** (XML in this case):

```
OrderId|PaymentType|OrderDate|Code,Qty,Price,|Name,Email,Phone,|Recipi
ent,Number,Street,City,State,Country,Pincode,|
```

```xml
<ns0:Order xmlns:ns0="http://BizTalkServicesOrderSample.OrderFF">
  <OrderId>MyOrder</OrderId>
  <PaymentType>CreditCard</PaymentType>
  <OrderDate>09-08-2013 22:50:00</OrderDate>
  <Product>
    <Code>100</Code>
    <Qty>1</Qty>
    <Price>500</Price>
  </Product>
  <Customer>
    <Name>Karthik</Name>
    <Email>user@hotmail.com</Email>
    <Phone>1-111-1111</Phone>
  </Customer>
  <ShippingAddress>
    <Recipient>Jon</Recipient>
    <Number>Building 1</Number>
    <Street>One Redmond Way</Street>
    <City>Redmond</City>
    <State>Washington</State>
    <Country>US</Country>
    <Pincode>98052</Pincode>
  </ShippingAddress>
</ns0:Order>
```

**Downloading the example code**

You can download the example code files for all Packt books you have purchased from your account at http://www.packtpub.com. If you purchased this book elsewhere, you can visit http://www.packtpub.com/support and register to have the files e-mailed directly to you.

# Creating the BizTalk Services solution

Open the bridge configuration surface (usually the `MessageFlowItinerary.bcs` file). The Visual Studio Toolbox should show the following entities:

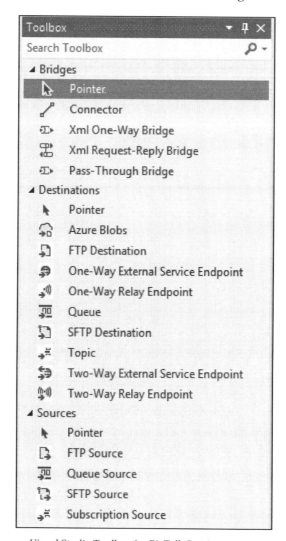

Visual Studio Toolbox for BizTalk Services project

Use the VS toolbox to drag-and-drop **FTP Source**, **Xml One-Way Bridge**, **One-Way Relay Endpoint**, and **Queue** and connect them using the **Connector**, as shown in the following figure:

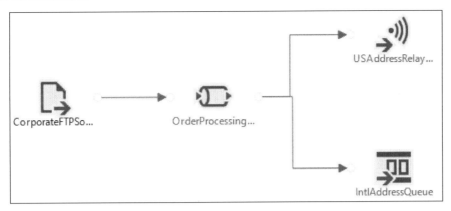

Bridge message flow in BizTalk Services project

Configure the following in the message flow:

1. Select the FTP server and configure the address, username, and password correctly.

2. Double-click on the bridge to open the **Xml One-Way Bridge** configuration:

   ○ In the **Message Types** block, add the `OrderFF.xsd` instance created earlier.

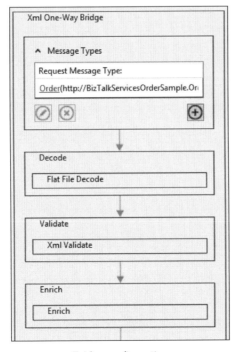

Bridge configuration

- ○ In the first Enrich stage, add an XPath Type property reading from `/*[local-name()='Order' and namespace-uri()='http://BizTalkServicesOrderSample.OrderFF']/*[local-name()='ShippingAddress' and namespace-uri()='']/*[local-name()='Country' and namespace-uri()='']` and writing to the location as a string. The XPath value can be obtained by opening the schema in VS, clicking on the relevant record, and copying the XPath value from the record properties window.

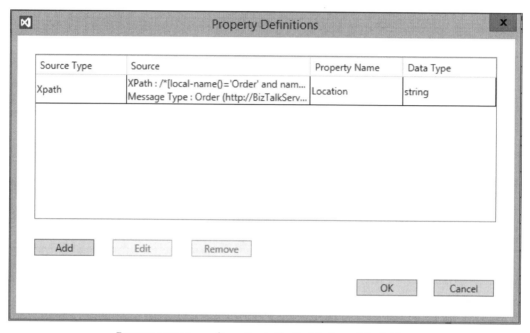

Promote property configuration in the Enrich stage of the bridge

3. In the parent `MessageFlowitinerary.bcs` view, click on the route link from **OrderProcessingBridge** to **USAddressRelay** and set the filter condition as **location='US'**; for the other link, set the location to US.

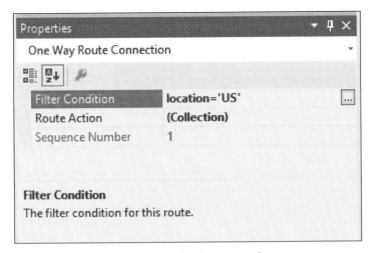

*Route properties for the message flow*

4. Edit the Queue .config file under MessageFlowitinerary.bcs and update the <tokenProvider> and the <endpoint> details with the Service Bus information.

5. Edit the Relay Service .config file under MessageFlowitinerary.bcs and update the <endpoint> details with the Service Bus relay information.

6. Build and deploy the solution.

7. If the deployment was successful, point your browser to https://<yourdeployment>/default/OrderProcessingBridge; you should see a 401 HTTP Error code stating a manage claim is required for this operation.

# Verifying the solution

We need to test sending two kinds of messages: one from the corporates and the other coming from the website:

- Load the MessageReceiver sample in VS and build the solution. From the output bin folder, run the following in a command prompt window:

  ```
  MessageReceiver.exe ServiceBusNS owner issuerkey USAddressRelay
  OneWayRelay
  ```

  Here, ServiceBusNS is the namespace where the relay is running and MyRelayTestSvc1 is the endpoint information configured in the bridge configuration.

- Load another MessageReceiver in a new command window.

  ```
  MessageReceiver.exe ServiceBusNS owner issuerkey IntlAddressQueue
  Queue
  ```

  Here, `ServiceBusNS` is the namespace where the Queue has been precreated and `IntlAddressQueue` is the endpoint information configured with the bridge.

- Load the MessageSender sample in VS and build the solution. `<yourdeployment>` is the URL where BizTalk Services was provisioned earlier.

  ```
  MessageSender.exe BizTalkSvcACS owner issuerkey
  https://<yourdeployment>/default/OrderProcessingBridge instance.
  xml application/xml
  ```

  Here, `BizTalkSvcACS` is the namespace of the BizTalk Service deployment ACS, `owner` and `issuerkey` are the ACS credentials of that namespace, and `instance.xml` is the `OrderFF.xsd` instance in XML format.

- The output is observed in the MessageReceiver of the relay.

- Edit `instance.xml` with `location=EU` and run the `MessageSender` command again. This time the output will be observed in the MessageReceiver of the Queue.

- Drop a flat file in FTP with `location=US` and observe the output in the relay service window.

- Drop a flat file in FTP with `location=EU` and observe the output in the message receive queue.

Northwind can now process both flat files and XML orders from either HTTP or FTP endpoints. You can delete the bridge from the BizTalk Services portal Bridge view or by using PowerShell.

# Summary

In this chapter, we have introduced the basics of BizTalk Services and the concepts, architecture, personas, and tools available to build an integration solution. We also exercised all the concepts learned through a simple order processing scenario with BizTalk Services and Service Bus relay and queues. The example can be further extended to include transforms, routing to other bridges like EDI, custom code, and so on. In the next chapter, we'll look at some of the BizTalk Services capabilities in more detail.

# 2
# Messages and Transforms

In *Chapter 1, Hello BizTalk Services*, we discussed the basics of **BizTalk Services** and the central concept of a bridge providing the vehicle for receiving and sending data via endpoint adapters (sources and destinations) through its built-in pipeline. In this chapter, we'll discuss messaging aspects, focusing on one particular aspect of messaging: transformation, or mapping. One of the most common aspects of integration is the need to turn one message format into another; what we referred to as structural impedance in *Chapter 1, Hello BizTalk Services*. It's the bread and butter of any integrator's toolbox, and BizTalk Services provides a brand new, modern mapping engine with graphical tooling to build sophisticated and powerful transforms. In this chapter, we'll look at BizTalk Service's mapping and transformation capabilities in detail and the flexibility it provides. To summarize, this chapter will cover the following:

- Why transformation and mapping is important
- Mapping capabilities in BizTalk Services
- Creating your first map
- Understanding mapping operations

## The problem

BizTalk Services' job is to let you connect this to that. What this and that actually are may not always be clear, well defined, or standardized into some internationally recognized protocol. A mapping capability is therefore crucial—a way to convert this into that. On many occasions, mapping requirements may be complex; the need to fundamentally change the shape or structure of a message, for example, or the need to replace data values from the source message with something that makes sense to the receiver. We can break this down into two classes of problems: one that needs to address the structure of a message, transformation; and one that needs to address its content, transcoding or translating it. The two types of mapping, transformation and translation, are both possible with BizTalk Services, as we'll see in this chapter.

# The mapper

So far, we're being deliberately vague, and with good reason. Often, mapping requirements are not well known and change as more is discovered about the nuances of the messaging formats involved and their variations. It often surprises people familiar with dealing with XML-based messages that describing their validity simply in terms of schema by using XML Schema Definition (http://www.w3.org/XML/Schema) can turn out to be more complicated than it first appears. This is unfortunately sometimes to do with the different productions or instances of XML messages that can be created or produced by a single schema, often unintentionally. XSD is sometimes not precise enough, and integration is therefore often messy, requiring good tools to make things fit, while the purity of standards and specifications doesn't go far enough to avoid ambiguity in implementation. This is a theme we'll come back to many times during this book: to be successful, any integration technology must be flexible to bend to the problem at hand, to fit into that, to not be changed, to adapt, to transform, and to integrate. Mapping is one tool in the box, and is a very important one to meet these requirements. As such, it deserves a chapter all to itself.

# The map designer

Take a look at the following screenshot. This shows the new graphical mapping designer that is accessible from Visual Studio 2012. For those familiar with **BizTalk Server**, don't be fooled. While it may have a similar look and feel to the BizTalk Server mapper, this tool has significant differences; the overriding design aesthetic was to make common mapping tasks as simple as possible, and as such, the mapper has undergone a significant overhaul.

Graphical mapping designer

# Schema

However, we're getting ahead of ourselves. In order to map one message format or structure to another, to translate its contents for example, we first need to understand the messages themselves. Fundamental to this is schema.

BizTalk Services differentiates between two types of messages: XML and non-XML. All XML message formats are expressed using XSD, and all non-XML message formats are expressed using XSD. So, XSD is important! The purpose of this book is not to provide a primer in XSD; we'll refer to other references if you need some background on the technologies we mention. Instead, we'll provide just enough to show how BizTalk Services uses such technologies so that the less familiar can still understand what is going on.

Now, you're probably wondering how any message formats you can think up can be defined in XSD. Let's look at an example.

# An example

Let's expand on the example we looked at in *Chapter 1*, *Hello BizTalk Services*. If you recall, this example received a file via SFTP and routed it to a Service Bus endpoint. Now we'll add a map to the solution. The map will transform the incoming message into a different format expected by the recipient. However, as noted previously, if we're to turn one message format into another, we need to define the schema of the target message first in order for us to be able to map to it.

To do this, right-click on the project, navigate to **Add | New Item**, select **Schema** from the list of items, and provide the name OrderUS.xsd. Click on **Add** to create the schema and add it to the solution.

The schema designer will now be open. As you did in *Chapter 1*, *Hello BizTalk Services*, add nodes to the schema to build it up, as shown in the following screenshot:

Changing the Order schema

Now, right-click on the project and navigate to **Add | New Item**. Select **Map** from the list of items and provide the name FFtoUS.trfm. Click on **Add** to create the map and add it to the solution.

The map designer will now open; the first task is to set the schemas. As a map's job is to convert one format to another, a minimum of two schemas are required: the input and the output.

Click on the **Open Source Schema** link, expand the tree, select OrderFF.xsd, and click on **OK**. Now click on the **Open Destination Schema** link, select the OrderUS.xsd schema, and click on **OK**.

The designer will now look like the following screenshot:

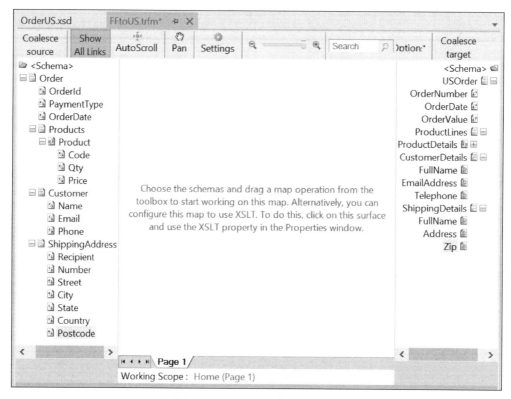

Selecting schema with the designer

Now we need to map one format to another. We do this by connecting the nodes together, usually working from left to right.

Join the `OrderId` node on the left-hand side to the `OrderNumber` node on the right-hand side by clicking and holding the left mouse button while dragging across to the right-hand side and releasing the button when the pointer is over the target field.

Now notice that the customer information is different in the target schema and the source; only the element names are different, but the structure is the same. The mapper provides a shortcut for mapping fields quickly to avoid having to connect them one by one. To do this, click and hold the left mouse button down on the parent `Customer` node in the left-hand side source schema and drag across to the target's `CustomerDetails` node. The context menu shown in the next screenshot will pop up. Here, we are presented with a number of options as shown in the following screenshot. Select **Link by Structure** and notice that all the nodes are connected together automatically even though their names differ. This is because this option connects fields in the order they appear, regardless of the node names, and is useful when the structure of both schemas is the same. You can use the same approach for mapping where the field names match too (**Link by Name**) or select **Simple Link** which will simply connect the top-level nodes together. This technique is very useful when mapping a large number of fields.

| Simple Link | |
|---|---|
| - | |
| Link by Structure. | |
| Link by Name | |
| Add mapeachs and direct links by matching names between selected source and target node hierarchies | |
| Cancel | |

Linking options

# Mapping operations

We can proceed like this for as many nodes as we like, connecting them individually or in groups. However, we often need to do more than just map one node's value to another. For this, we can turn to mapping operations. For those acquainted with BizTalk Server, you will be familiar with **functoids**; the concept is similar in BizTalk Services. However, despite the similarities, there are many differences in how they are realized. One of the primary goals of the product group is to simplify common tasks, such as looping, which were often difficult or time consuming to achieve previously. This is where we will now focus our attention.

BizTalk Services provides a total of 37 mapping operations that are functionally grouped into categories in the toolbox. There isn't room here to cover every mapping operation, so we'll focus on some of the most useful. For a complete reference, check the MSDN documentation at `http://msdn.microsoft.com/en-us/library/windowsazure/hh689870.aspx`. The idea is that all mapping operations are configured and connected in the same way; so, once you've learnt the operations available, it becomes a straightforward case of using a combination of them to get the mapping job done. The mapping operation categories are listed in the following table:

| Category | Purpose |
| --- | --- |
| String operations | Manipulate node values as strings such as concatenation, trimming, and substring operations |
| Loop operations | Operations to loop round repeating nodes in source |
| Expressions | Arithmetic and logical expressions to perform calculations or decisions |
| List operations | Processing for lists of items that can be created from node content in a conditional way for further processing |
| Cumulative operations | Operations to accumulate values such as sums, counts, and averages |
| Date and time operations | Manipulate date and time values |
| Miscellaneous operations | Various operations for retrieving context properties, formatting numbers, and incorporating C# in your maps |

A very common type of transformation is flattening. This is where a number of repeating items (usually a list) needs to be consolidated (or flattened) into a single value, often with some calculation applied (for example, a summation). BizTalk Services provides several mapping operations to achieve this in a straightforward way.

Take a look at the following XML and you can see that the `<Product>` element repeats, that is, there can be more than one product specified. Let's say we want to calculate the sum of all the product prices (`Price`) multiplied by the quantity (`Qty`) ordered to work out the total value of the order:

```
<ns0:Order xmlns:ns0="http://BizTalkServicesOrderSample.OrderFF">
  <OrderId>OrderId_0</OrderId>
  <PaymentType>PaymentType_0</PaymentType>
  <OrderDate>OrderDate_0</OrderDate>
  <Products>
    <Product>
```

```
      <Code>Code_0</Code>
      <Qty>Qty_0</Qty>
      <Price>Price_0</Price>
    </Product>
    <Product>
      <Code>Code_0</Code>
      <Qty>Qty_0</Qty>
      <Price>Price_0</Price>
    </Product>
    <Product>
      <Code>Code_0</Code>
      <Qty>Qty_0</Qty>
      <Price>Price_0</Price>
    </Product>
  </Products>
  <Customer>
    <Name>Name_0</Name>
    <Email>Email_0</Email>
    <Phone>Phone_0</Phone>
  </Customer>
  <ShippingAddress>
    <Recipient>Recipient_0</Recipient>
    <Number>Number_0</Number>
    <Street>Street_0</Street>
    <City>City_0</City>
    <State>State_0</State>
    <Country>Country_0</Country>
    <Postcode>Postcode_0</Postcode>
  </ShippingAddress>
</ns0:Order>
```

The final result, the total value of the order, then needs to be mapped to a single field in the target schema. We can use mapping operations to easily achieve this. The key to using mapping operations successfully is to break down the requirements and select the most appropriate mapping operations to achieve the goal. As my explanation has already hinted at, there are several parts to this problem. The first of these is to realize that we need to keep a running total of each of the `Qty * Price` field calculations—one for each `Product` element. Let's deal with that first.

BizTalk Services provides a set of list-based mapping operations that allow the creation of a temporary list to store items and manipulate the items in the list.

First drag a **Create List** mapping operation onto the design surface. Perhaps the first obvious change to a BizTalk Server developer is that mapping operations in BizTalk Services provide nesting. This is the key to simplifying complex tasks as this nesting behavior provides a natural way to group and organize the mapping task required.

The **Create List** operation will be used to hold the temporary results from our calculations. We will push each product's total to the list and then calculate the sum of these list item values later. Double-click on **Create List** to configure it. In the dialog, type a **Member Name**, say total, and select **Number** from the dropdown for the **Member Type**, as shown in the following screenshot. This is a variable that will be used to store the value of our calculations. Click on **OK** to close the dialog.

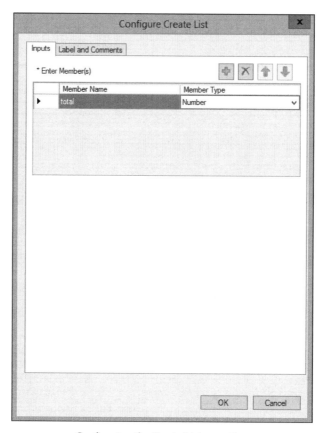

Configuring the Create List operation

The next step is to loop round all the Product elements. To do this, drag a **ForEach** mapping operation and drop it inside the **Create List** operation. Notice how we can place additional operations inside this operation too. This is where we'll put the calculation of each line item's total.

Now wire up the `Product` node in the left-hand schema to the **ForEach** operation. This tells the operation to loop round each `Product` node within `Products`.

Drag an **Arithmetic Expression** operation across to the designer and drop it in the **ForEach** operation. Now wire up the `Qty` and `Price` fields to this operation. These will be our input parameters; the nodes from the input message we want to use the data from. Taking each node in turn, drag a connection onto the **Arithmetic Expression** operation on the canvas. Now double-click on the new **Arithmetic Expression** to configure it. Here we can specify a calculation based on the fields that are connected to the operation, in this case `Qty` and `Price`. Enter the expression shown in the following screenshot:

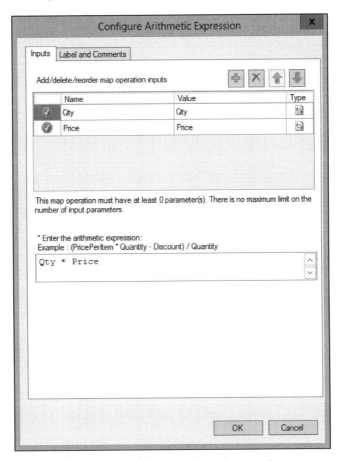

Configuring the Arithmetic Expression operation

We now need to store this result; to do this, we'll add it to the outer list operation. Drag an **Add Item to List** operation and drop it to the right of the **Arithmetic Expression** operation, within the **ForEach** operation. Then, connect the **Arithmetic Expression** operation to the **Add Item to List** operation by dragging a line from one to the other. Double-click on the **Add Item to List** operation to configure it. The dialog should be pre-populated already, so we can just click on **OK** to save the settings. The map should now look as shown in the following screenshot:

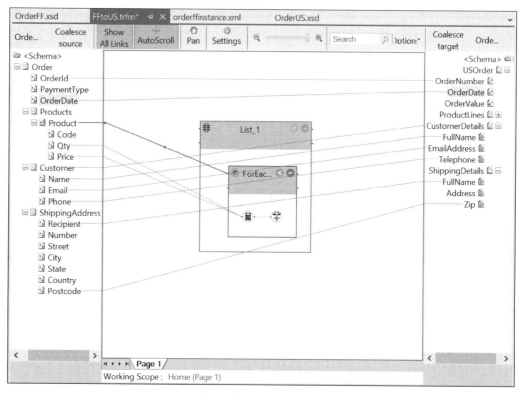

Partially-completed map

OK, the first part of the task is done; we are calculating the total value for each product. We now need to add these totals together to obtain the grand total for the order. This is very straightforward. Drag a **Select Entries** mapping operation and drop it to the right of the **Create List** operation. Connect the **Create List** operation to **Select Entries**. Double-click on **Select Entries** to open it, check the **Selected** checkbox next to `total`, and click on **OK** as shown in the following screenshot. Here, we are specifying what variables we want to extract from the list we've created. As we only have one in this case, the choice is easy.

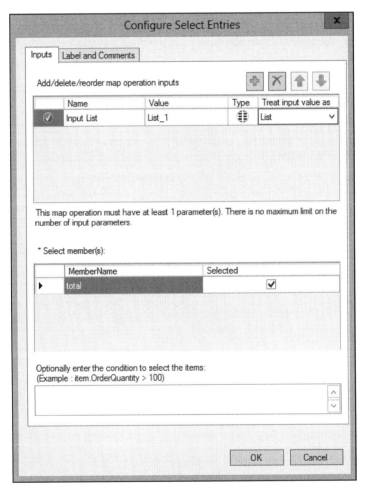

Configuration of the Select Entries operation

Finally, drag a **Cumulative Sum** operation and drop it to the right of the **Select Entries** operation. Connect the **Select Entries** operation to the **Cumulative Sum** operation. Now connect the **Cumulative Sum** operation to the `TotalValue` field in the target schema. Double-click on the **Cumulative Sum** operation you just added and enter `item.total` in the expression textbox, as shown in the following screenshot. Here, we are specifying our total variable from the list entries passed from the **Select Entries** operation:

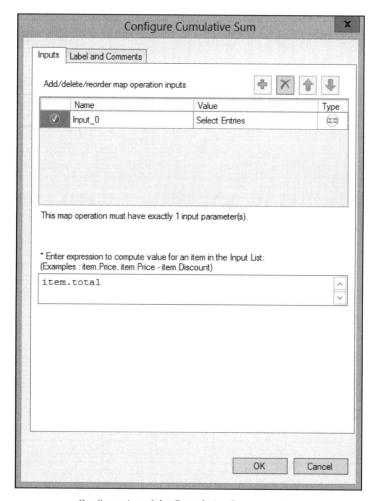

Configuration of the Cumulative Sum operation

The completed map should look the same as the following screenshot:

Map with looping and calculated node

We're nearly there with our map. Let's tackle the `ShippingAddress` node. Notice that there are more fields on the left-hand side than the right. We're therefore going to consolidate these by concatenating some of them. There are a few ways to do this; we could use the **String Concatenate** mapping operation, for example, that can take any number of inputs and produce a concatenated single string as output.

However, let's look at something a little more interesting. When there isn't an operation that meets your need, you can turn to the **CSharp Scripting** operation. As its name implies, this operation allows you to include C# in its configuration, giving you the full power of the .NET framework to be able to implement whatever mapping functionality you need.

Drag a **CSharp Scripting** operation to the mapping design surface and drop it somewhere below the **Create List** operation. Connect the `Number`, `Street`, `City`, `State`, and `Country` nodes to it. As you've probably realized by now, this is the way operations can work on specific data items, and the same is true for the scripting operation. By connecting these items to it, they become available to the scripts we write. Double-click on the **CSharp Scripting** operation to open its configuration.

In the dialog, notice that there is a **Script Text** multiline textbox. Here, we can define a function in C# that will take the nodes as input parameters and return a value. In this simple case, you can see from the next screenshot that I am just concatenating the input parameters with some formatting and returning the result back to the map. Enter the following code to do the same:

```
public string CreateAddress (string Number, string Street, string
City, string State, string Country)
{
    return Number + " " +
           Street + "," +
           City + "," +
           State + "," +
           Country;
}
```

Note that the names of the input nodes to the scripting operation must match the argument names in the code. If the names differ, the map won't compile. Once we've saved this by clicking on **OK**, we can connect the operation to the target schema's `Address` node.

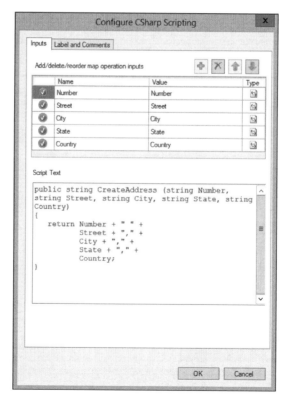

Using C# with mapping

There are only a few fields left to map now. Simply connect `Recipient` to `FullName` and `PostCode` to `Zip`. The final operation we'll look at to complete this map is the **DateTime Reformat** operation. Drag this onto the design surface above the **Create List** operation. Connect the `OrderDate` node from both schemas to this operation, then double-click on the **DateTime Reformat** operation to configure it. This operation is useful when dealing with date formatting requirements that differ between sender and receiver. The nice thing about this operation is that it doesn't only support a fixed set of date and time formats, but you can enter your own as well. For the **Input Format** field, enter `d/M/yyyy` in the **Format** field. Note that this is not one of the provided options in the dropdown, so you'll need to enter it in the textbox, as shown in the following screenshot. Also, make sure that the letter `M` is capitalized, as is shown. Then, select **M/d/yyyy** for the **Output Format** field. This will interpret the input date in day/month/year format, for example, `2/9/2013`, and change to month/day/year format for the output, for example, `9/2/2013`.

Formatting dates

The map is now complete. It should look similar to the following screenshot:

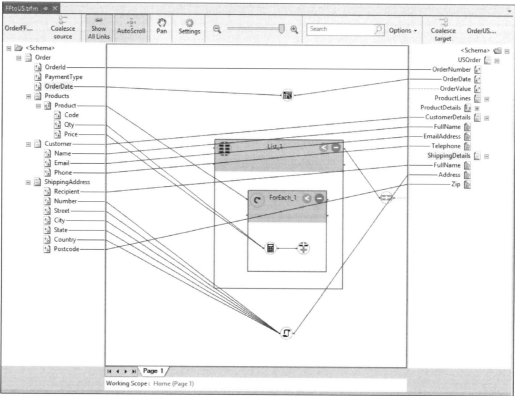

The completed map

# Testing

Phew! This may seem quite complex, but it's really quite simple once you break it down. The next step is to test the map and see if it looks like it's producing the right results. Testing is built right into Visual Studio, so there's no need to compile and deploy the solution to Windows Azure in order to see if it works. This is important as creating anything other than trivial maps is a very iterative process. It's made easier by building up the functionality in the map gradually and examining the test results along the way. This way, any mistakes are much more obvious and easily corrected.

To test a map, we need some input. This is most easily generated in Visual Studio itself. Right-click on the `OrderFF.xsd` schema in the **Solution Explorer** window and select **Generate Instance**. Open the file that's created and edit the values to match the one shown in the following code (don't forget, you can just download the source for this example from the website):

```
<ns0:Order xmlns:ns0="http://BizTalkServicesOrderSample.OrderFF">
  <OrderId>123</OrderId>
  <PaymentType>ACCOUNT</PaymentType>
  <OrderDate>2/9/2013</OrderDate>
  <Products>
    <Product>
      <Code>AB12</Code>
      <Qty>4</Qty>
      <Price>1.50</Price>
    </Product>
    <Product>
      <Code>AC01</Code>
      <Qty>2</Qty>
      <Price>3.99</Price>
    </Product>
    <Product>
      <Code>DE4</Code>
      <Qty>10</Qty>
      <Price>12.25</Price>
    </Product>
  </Products>
  <Customer>
    <Name>John Doe</Name>
    <Email>john.doe@contoso.com</Email>
    <Phone>425-123456</Phone>
  </Customer>
  <ShippingAddress>
    <Recipient>Jane Smith</Recipient>
    <Number>1</Number>
    <Street>East Street</Street>
    <City>New York</City>
    <State>New York</State>
    <Country>USA</Country>
    <Postcode>NY12345</Postcode>
  </ShippingAddress>
</ns0:Order>
```

The **Generate Instance** action creates an XML-format message by default — which is what you need for the map itself. However, this schema is a flat file schema, and if we want to generate a message to pass into the bridge, we need to generate a message in this format. In the schema properties, there is a property called **Generate Instance Output Type** that can be set to **Native** instead of **XML**. When **Native** is selected, the schema will create a test message according to its type, flat file or XML. The following screenshot shows the results of setting this to **Native** when compared with the OrderFF.xsd schema:

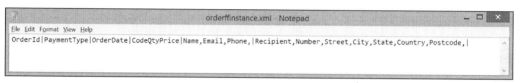

Generate Instance: Native

Once we have a test message, we can assign it to the map to try it. Click on the FFtoUS.trfm map in the **Solution Explorer** window; in the **Properties** window, enter the path to the file in the **Test Map Input File** property. Now right-click on the map and select **Test Map**. With any luck, you should see something similar to the following snippet in the output window. This means that the map execution worked!

```
Test Map succeeded.
Output is written to the file 'C:\Users\Jon\AppData\Local\Temp\
tmp3817.xml'.
```

Open the file from the **File** menu by navigating to **Open | File** and browsing to the to the file location in the preceding output (in this case: C:\Users\Jon\AppData\Local\Temp\tmp3817.xml). If you did everything right, it should look the same as the following XML code:

```
<?xml version="1.0" encoding="utf-8"?>
<ns1:USOrder xmlns:ns0="http://BizTalkServicesOrderSample.OrderFF"
xmlns:ns1="http://BizTalkServicesOrderSample.OrderUS">
  <OrderNumber>123</OrderNumber>
  <OrderDate>9/2/2013</OrderDate>
  <OrderValue>136.48</OrderValue>
  <CustomerDetails FullName="John Doe" EmailAddress="john.doe@contoso.
com" Telephone="425-123456">
  </CustomerDetails>
  <ShippingDetails FullName="Jane Smith" Address="1 East Street,New
York,New York,USA" Zip="NY12345" />
</ns1:USOrder>
```

Notice how different this XML document is from the one you used as input and you can hopefully start to appreciate the power of the BizTalk Services mapper.

# Configuring a bridge

A map is no good on its own though. We need to be able to use it in an integration solution. It should hopefully come as no surprise that the way we do this is by configuring a bridge. The next screenshot shows part of the bridge configuration. This configuration represents the pipeline of processing that can be configured. There are multiple stages to this pipeline, as mentioned in *Chapter 1, Hello BizTalk Services*. In the middle of the pipeline, there is a **Transform** stage; it is here that we can specify a map to execute.

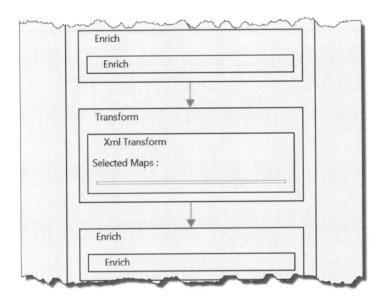

Configuring a bridge with a map

Double-click on the `MessageFlowItinerary.bcs` file in the **Solution Explorer** window to open it. In the designer, open the `OrderProcessing` bridge configuration by double-clicking on it. Click on the **Transform** stage and look at the **Properties** window. Here, we can choose a map by clicking on the ellipsis (**...**) next to the **Maps** property to open up the configured maps. This will show us all the maps the solution contains.

We can select the map that we created earlier by checking the **Selected** checkbox next to it, as shown in the following screenshot. Clicking on **OK** will return to the bridge configuration, which should now show the selected map, FFtoUS, in the **Transform** stage.

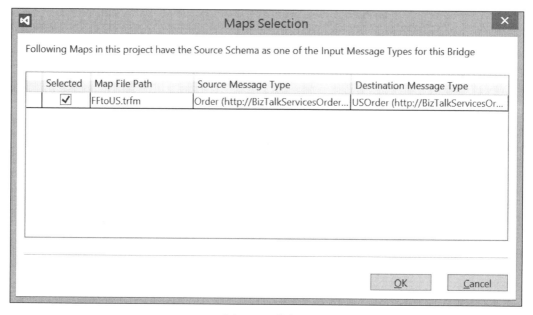

Select map dialog

# Putting it all together

The solution is now ready. Build and deploy as before, and once deployed, point your browser to `https://<yourdeployment>/default/OrderProcessingBridge` and you should see a 401 HTTP error code stating that a manage claim is required for this operation.

Now you will use two tools provided as a part of the BizTalk Services SDK. These are MessageSender and MessagerReceiver which you can download from the following links. These tools allow you to send messages to and receive messages from the bridge you created:

- `http://code.msdn.microsoft.com/windowsazure/Windows-Azure-BizTalk-EAI-e01a5b64`

- `http://code.msdn.microsoft.com/windowsazure/Windows-Azure-BizTalk-EAI-af9bc99f`

Unzip both solutions and open the MessageReceiver sample in Visual Studio 2012 and build it. Run it from the command prompt by typing the following and pressing the *Enter* key.

```
<path>MessageReceiver.exe ServiceBusNS owner <issuerkey> USAddressRelay
OneWayRelay
```

In the preceding command, `<path>` is the path to the exe from the build output of Visual Studio, `ServiceBusNS` is the namespace where the relay is running and `USAddressRelay` is the endpoint information configured in the bridge configuration. Note that you will also need to replace the `<issuerkey>` value with your own subscription details.

Now open the MessageSender sample (downloaded from the previous link) in Visual Studio 2012 and build it. Run it as shown in the following code to send a message to the bridge:

```
<path>MessageSender.exe BizTalkSvcACS owner issuerkey
https://<yourdeployment>/default/OrderProcessingBridge instance.xml
application/xml
```

In the preceding code, `BizTalkSvcACS` is the ACS namespace of the BizTalk Service deployment. As before, `owner` and `issuerkey` are the ACS credentials of that namespace, and `instance.xml` is the `OrderFF.xsd` instance in XML format.

Ensure that output is observed in the MessageReceiver sample of the relay. Examine the output message and notice how the map has transformed it.

# More on mapping

We've covered a lot of ground so far, but there is much more to mapping in BizTalk Services besides the other 27 operations we've not used here. There are two other groups of operations that deserve some discussion.

The first of these is the **Get Context Property** mapping operation. This provides an often-asked-for feature in BizTalk Server—the ability to retrieve properties from the message context and include them in a map. The way this works is by configuring it, specifying the property name to retrieve it, and then connecting it to a target node; no input nodes are required. We haven't covered context properties in detail yet, but for now, remember that they are a set of name/value pairs that hold contextual information about the current message flow; for example, the transport details (for example, a filename) of the message received, or even properties of the message itself that have been extracted. If you're wondering how you can test this from within Visual Studio as we did earlier, the team has thought of this too. A property, **Context Property Test Data**, is provided on the map and allows you to specify the test name/value properties to execute the map with. The dialog is shown in the following screenshot. The ability to use context properties in BizTalk Services maps is a very welcome addition. This dialog can also be shown when the map is tested to change the values used.

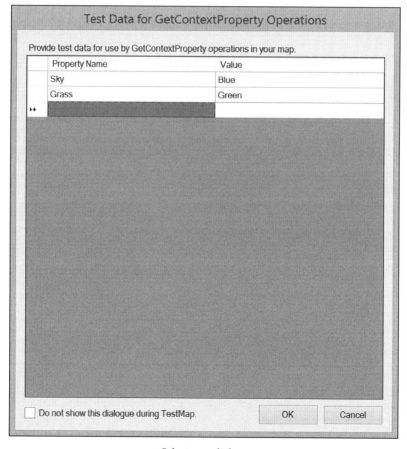

Select map dialog

The second area of improvement is in expression operations. For example, an **If-Then-Else Expression** operation is provided. This greatly simplifies the common requirement of testing a condition; if it evaluates to true, one path is taken, otherwise another. In BizTalk Server, this was complex to achieve, requiring a lot of functoids. This highlights the effort that the product group has put in here to simplify common tasks, as I mentioned at the beginning of this chapter. The same goes for other logical operations such as the **Logical Expression**. Here, an expression can be provided that evaluates to true or false. Again, for those familiar with BizTalk Server, this operation replaces a large number of functoids with just one that is simple to configure and use.

One issue that was starting to become obvious, referring to the map in the earlier screenshot, is complexity. The mapper provides the concept of pages to allow the splitting up of operations and links on different pages. You can add a new page by right-clicking on the area next to the tab at the bottom of the designer, as shown in the following screenshot:

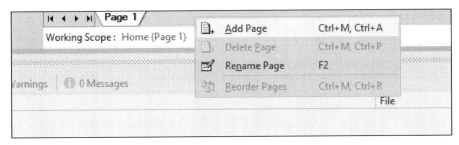

Working with pages

Splitting the map up into different pages is a trade-off between readability and complexity. Ideally, you would see as much detail as is clearly readable in a single page to avoid having to jump between different pages all the time. With complex maps that perhaps have thousands of links and operations, this isn't possible; adding pages can greatly reduce complexity and improve clarity, especially for those left to maintain the solution.

One final point is that, as you may have already noticed, the new BizTalk Services mapper is not based on the Extensible Stylesheet Language and XSLT (as the BizTalk Server one was). However, it is still possible to include XSLT (1.0 only) in a map, which is useful when you have an existing transformation in XSLT that you wish to reuse. The XSLT property is accessed by clicking on the designer grid and opening the **XSLT** property in the **Properties** window. On the subject of reuse, another useful tool provided with BizTalk Services is the BizTalk Server map converter. This will convert a BizTalk Server .btm map file to the BizTalk Services mapping format, saving time when you have existing maps that you wish to reuse and avoiding the need to start from scratch. Because of the differences in functionality between the two, it cannot perform a 100 percent conversion, but is a great time-saver nevertheless.

# Dealing with failure

One very important point a developer must consider is how to deal with failures that occur. In integration solutions, failure is particularly important as it can be hard to isolate and diagnose. On a map, it is possible to configure what action to take should a particular operation fail, usually due to bad data provided to it. The **Settings** button at the top of the designer displays the dialog, as shown in the following screenshot, when clicked.

Here, each operation (or group of operations in some cases) can be set to either **Fail Map** on an error occurrence or continue and output a null value. This is very useful; we'll look at error handling in much more detail in a later chapter.

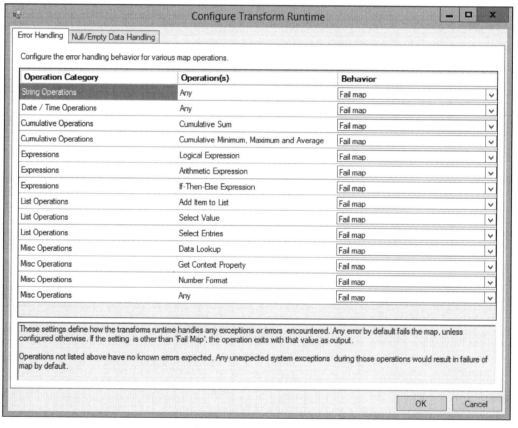

Setting runtime properties

# Summary

In this chapter, we have looked at the mapping features of BizTalk Services. You've seen how to create maps, use many of the powerful operations provided, and test them. While we haven't been able to cover every operation, many are self-explanatory and easy to understand; after all, the whole point of the mapper is to make the job of format and content conversion easier. We urge you to experiment on your own and see what you can come up with.

In the next chapter, we'll look at the different types of bridges BizTalk Services provides, starting with EDI.

# 3
# Bridges

In *Chapter 2, Messages and Transforms*, we covered a fundamental aspect of integration and transformation. But transformation is just one of the capabilities that bridges in BizTalk Services provide. In this chapter, we will take a closer look at bridges and the following features:

- Pipeline stages
- Validating, enriching, and formatting messages
- Lookup data
- Message routing and filters
- BizTalk Services Explorer

A **bridge** is actually a **Windows Workflow Foundation (WF4)** behind the scenes. While you cannot create your own bridge definitions, three templates are provided for you:

| Type | Description |
|------|-------------|
| **XML One-Way** | Caller sends XML-based messages to the bridge and expects no response |
| **XML Request-Reply** | Caller sends XML-based messages and waits for response message |
| **Pass-Through** | Caller sends message in any format (XML or non-XML) in a one-way pattern |

These templates provide some standard processing steps that you can use to act on or affect messages as they are processed. These steps form a pipeline of processing, with each step following the previous one sequentially. Each step also acts on the state of the message and its context from the previous step as well. There are also opportunities for you to add your own custom pipeline processing as we'll see later in this chapter.

It is worth remembering that bridges are inherently stateless—there is nothing durably persisted during bridge processing. If a bridge fails during processing, the message could be lost, and so care must be taken to avoid this situation. We'll come back to this in much more detail later in this book in *Chapter 7, Tracking and Troubleshooting*.

# Pipeline processing

Within a bridge's pipeline, there are the following steps:

| Stage | Direction | Purpose |
| --- | --- | --- |
| Message Type | Receive | Match schema to incoming message |
| Decode | Receive | Convert incoming message to XML based on schema |
| Validate | Receive | Determine if message is valid according to the schema |
| Enrich | Receive | Create properties from message or context content |
| Transform | Receive/Send | Map the message to another message schema format |
| Enrich | Send | Create properties from the message or context content |
| Encode | Send | Get the message ready for transmission |

Of course, for two-way and pass-through bridges, things are a little different, as you would expect. For two-way bridges, there are no Decode and Encode stages, and pass-through bridges only have a single stage, Enrich.

# Message processing

As BizTalk Services hosted in the cloud exposes default HTTP endpoints to the bridges that you publish, this means that it is possible to submit messages to a bridge by simply posting them to the endpoint and, with the request/reply bridge, receive a response.

Of course, BizTalk Services provides many more message sources and destinations such as FTP, Service Bus queues and topics, and also line of business systems such as SAP that are covered in detail in *Chapter 4, Enterprise Application Integration*. The full list is provided in the following table:

| Transport | Source | Destination | Description |
|---|---|---|---|
| FTP | Yes | Yes | File Transfer Protocol support |
| SFTP | Yes | Yes | Secure File Transfer Protocol |
| Service Bus Queue | Yes | Yes | Receive and send messages to/from queues |
| Service Bus Topic | Yes | Yes | Receive and send messages to/from topics |
| HTTP(S) | Yes | Yes | Bridges are exposed as HTTPS endpoints by default on the namespace you create the service under. BizTalk Services also supports HTTP as a destination by allowing the calling of web services. |
| Azure Blob Storage | No | Yes | Send a message to Azure Blob Storage |
| Relay | No | Yes | The Service Bus relay is used with BizTalk Adapter Services to connect to the following line of business systems:<br><br>• SQL Server<br>• Oracle DB<br>• Oracle eBusiness Suite<br>• mySAP<br>• Siebel<br><br>This is covered in detail in *Chapter 4, Enterprise Application Integration*. |

# Messaging

With XML bridges, there are two stages that are used to identify received messages and determine what to do with those that are not expected. In order to know what's not expected, the Message Types section of the bridge allows the specification of message schemas. Any number of schemas can be configured under the **Message Type Picker** dialog box as shown in the following screenshot:

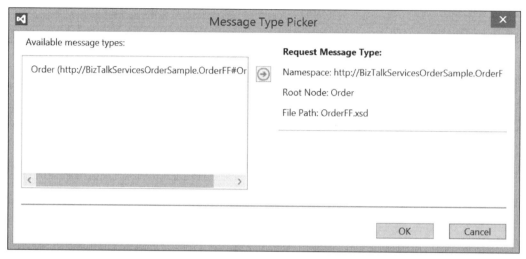

Message Types

The first stage in an XML bridge is Decode. This is only applicable for flat files, that is, messages that are not received as XML and have a flat file schema specified (more on flat file schemas in *Chapter 4*, *Enterprise Application Integration*). While processing flat files in an XML bridge may sound strange, the purpose of the bridge is to normalize the message into an XML format as this is what allows uniformity of processing useful features such as transformation and enrichment. In this way, data in any format, for example JSON, can be received and processed. There is nothing to configure in the Decode stage; instead, it takes its configuration from the provided message types and applies the matching flat file schema to create an XML representation of the file. The match is made based on the schemas selected in the **Message Type Picker** dialog box as shown in the preceding figure. Only one message type can be chosen.

After the Decode stage, the message is validated against the schemas provided. The only configurable property here is the Boolean setting **Report Warnings As Errors**. This defaults to `false`, meaning that unrecognized or invalid messages are still processed through the rest of the bridge. Setting this property to `true` will throw an error in the bridge and the message will not be processed. The caller (if the caller is using HTTP) will receive an **HTTP 500** status code response. This general "Server Error" response is generally returned with a response detailing the problem and providing a tracking ID that can be used to diagnose the cause. Fault diagnosis and troubleshooting is covered in more detail in *Chapter 7, Tracking and Troubleshooting*. If FTP is the configured source, then the file is left on the FTP server and the bridge will be retried up to three times after waiting a number of minutes (which extends over the number of retries). This behavior is not currently configurable in WABS.

# Enrichment

Enrichment occurs at two points in the bridge: pre and post transformation. The enrichment stages provide the opportunity to write to message properties that can be used in either transformation (in the first Enrich stage) or in routing (post transformation). Message properties are simply name/value pairs that are moved through the bridge with the message itself and can be created in the Enrich stages.

There are several sources of data available when writing to a message property, and these are listed in the following table:

| Source type | Purpose |
| --- | --- |
| Soap | Access SOAP properties of the message such as the Action |
| Http | Access HTTP headers sent by the caller |
| Lookup | Look up a value in a Windows Azure SQL database |
| Xpath | Look up a value using an XPath expression in the message |
| Ftp | Access FTP properties such as filename if source is FTP |
| Sftp | Access SFTP properties if source is SFTP |
| System | Provides access to the system properties such as the date/time a message was received |
| Brokered | Access Service Bus properties if source or destination is queue/topic based |

Using message properties allows bridge processing to be influenced and controlled in two primary ways:

- **Through transformation**: The message received from the caller can be transformed into a different format that is required by the final receiver. Message properties can be accessed.

- **Through routing**: We'll look at this in detail in a moment. Message properties can be used to direct a message to a particular destination.

Let's look at a couple of examples. In the following figure, we have a bridge with two destination Service Bus queues configured, **Europe** and **Americas**. Assume that we would like to create a property that holds the value of a field in the incoming message that contains the name of the country the message is from. The routing bridge pattern is shown in the following screenshot:

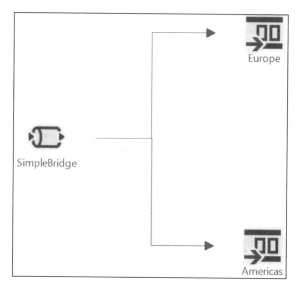

Routing bridge pattern

The incoming message looks like the following XML:

```
<ns0:Order xmlns:ns0="http://BizTalkServicesOrderSample.Order">
  <OrderId>123</OrderId>
  <PaymentType>ACCOUNT</PaymentType>
  <OrderDate>2/9/2013</OrderDate>
  <Products>
    <Product>
      <Code>AB12</Code>
      <Qty>4</Qty>
      <Price>1.50</Price>
    </Product>
    <Product>
      <Code>AC01</Code>
      <Qty>2</Qty>
      <Price>3.99</Price>
    </Product>
    <Product>
      <Code>DE4</Code>
      <Qty>10</Qty>
      <Price>12.25</Price>
    </Product>
  </Products>
  <Customer>
    <Name>John Doe</Name>
    <Email>john.doe@contoso.com</Email>
    <Phone>425-123456</Phone>
  </Customer>
  <ShippingAddress>
    <Recipient>Jane Smith</Recipient>
    <Number>1</Number>
    <Street>East Street</Street>
    <City>New York</City>
    <State>New York</State>
    <Country>UK</Country>
    <Postcode>NY12345</Postcode>
  </ShippingAddress>
</ns0:Order>
```

We can create an XPath property against this message as follows:

1. Click on the first Enrich stage in the designer.

2. In the **Properties** window, double-click on the **Property Definitions** collection ellipsis.

3. In the **Property Definitions** dialog box, click on **Add**.

4. Select a type of XPath.

5. Enter the following expression in the **Identifier** field:

```
/*[local-name()='Order' and namespace-
  uri()='http://BizTalkServicesOrderSample.Order']/*[local-
  name()='ShippingAddress' and namespace-uri()='']/*[local-
  name()='Country' and namespace-uri()='']
```

> To easily get the XPath of an item from an XML schema, open the `.xsd` file in the Schema editor, select the item you want, and then look in the **Properties** window. The **Instance XPath** property will contain the XPath expression required to extract it at runtime.

6. Specify the message type of the `Order` instance.

7. Enter `MappedCountry` in the **Property Name** field.

8. Select **string** for the **Data Type** field.

9. Click on **OK**.

The **Property Definitions** dialog box should now look like the following screenshot. At runtime, when messages are received by the bridge, the Enrich stage will extract the field from the message using the specified XPath and store the country value in the message property.

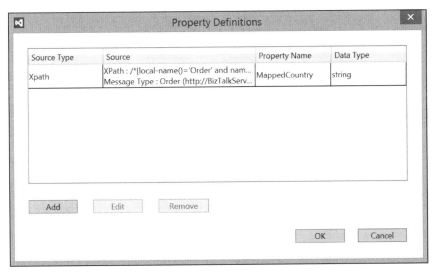

| Source Type | Source | Property Name | Data Type |
|---|---|---|---|
| Xpath | XPath : /*[local-name()='Order' and nam...<br>Message Type : Order (http://BizTalkServ... | MappedCountry | string |

Add    Edit    Remove

OK    Cancel

Property Definitions

# Lookups

Another usage of message properties is with transformation. Here, a common requirement is transcoding, where one value needs to be replaced with or mapped to another value. Code tables can be used for this purpose, and we can use the lookup capabilities of BizTalk Services to do this and then feed the values into a transform to replace the value in the message.

Some preparation is needed to set things up. If you recall, when you provision a new Windows Azure BizTalk Services instance, you can choose to create a new SQL Azure database to hold the various tables needed by the service. We can also create a table in this database to hold the transcoding data for the lookups. To create a table and add data to it, run the following script against the database:

```
CREATE TABLE [dbo].[CountryMap](
  [CountryName] [nvarchar](100),
  [ISOCountry] [int] NULL
  CONSTRAINT [PK_CountryName] PRIMARY KEY CLUSTERED
  (
    [CountryName] ASC
  ) WITH (PAD_INDEX = OFF, STATISTICS_NORECOMPUTE = OFF,
    IGNORE_DUP_KEY = OFF, ALLOW_ROW_LOCKS = ON, ALLOW_PAGE_LOCKS =
    ON)
)
GO
INSERT INTO [dbo].[CountryMap] (CountryName, ISOCountry)
  VALUES('USA',844),('UK',826),('CANADA',124)
```

The easiest way to do this is to go to the Azure Management Portal at
http://manage.windowsazure.com and click on the **SQL Databases** tab.
The database created for your BizTalk Services instance will be named with
the service name you provided, appended with the _db extension. Click on this
database and then click on **Manage**. A new window (or tab) will open in the
browser as shown in the following screenshot. In this window, you can select
**New Query**, paste in the preceding SQL code, and click on **Run** to create the table
and populate it with some data.

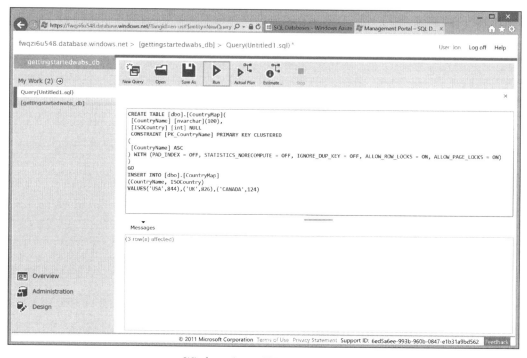

Windows Azure SQL query editor

As shown in the dialog box of **Property Definitions**, we can configure the Enrich
stage in the bridge as before by performing the following steps:

1. Open the **Property Definitions** dialog box on the first Enrich stage.

2. Select **Lookup** for the **Type** field.

3. Click on the dropdown list for the **Identifier** field and click on
   **Configure New...**.

4. The dialog box will be displayed; complete it as shown in the following screenshot:

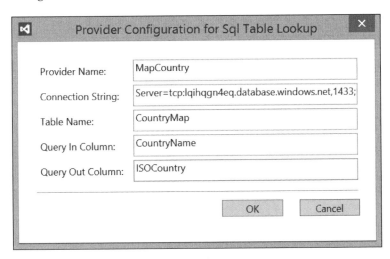

Provider configuration

> To get the **Connection String** value, log back into the Azure Management Portal, click on the **SQL Databases** tab as you did earlier, and click on the WABS database that was created when you provisioned your service. Click on **Dashboard** and then click on **Show Connection Strings**. Copy the value in the ADO.NET textbox into this field.

5. Click on **OK** to close the dialog box.

6. For the **Lookup** property, select **MappedCountry** — this is the context property that was created by the XPath earlier and is used as input to the lookup.

7. Enter `MappedCountry` in the **Property Name** field.

8. Select **string** in the **Data Type** dropdown.

9. Check the values as shown in the following screenshot:

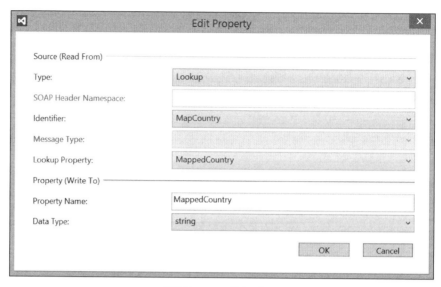

Edit Property dialog box

10. Click on **OK** to create the property.

11. Finally, click on **OK** to close the **Property Definitions** dialog box.

Now, when a message is received, the country name in the message is looked up in the database and the ISO country code will be returned and stored in the **MappedCountry** property. To finish off, we need to add a transformation to update the message itself with the following steps:

1. Right-click on the project and select **Add | New Item...**.

2. Select **Map** in the templates dialog box and provide a map name of `CountryNameToCountryCode.trfm`.

3. Click on **OK** to create the map.

4. In the open map, click on the **Open Source Schema** link. Select the **PO.XSD** schema (from *Chapter 2, Messages and Transforms*).

5. Do the same for the **Open Destination Schema** link.

6. In the functoids toolbox, drag-and-drop a **Get Context Property** functoid to the designer (it's located in the **Misc Operations** section).

7. Double-click on the functoid on the map to configure it.

8. In the **Property Name** field, enter `MappedCountry`.

9. Click on **OK** to close the dialog box.

10. Hold down the *Shift* key and click and hold the left mouse button on the
    `Order` node in the left-hand side schema. While still holding the left mouse
    button and the *Shift* key, drag across to the `Order` node in the right-hand side
    schema to connect them. In the pop up that is shown, select **Link by Name**.
    Recall from *Chapter 2, Messages and Transforms*, that this action will map every
    field from the source to the destination.

11. Finally, you need to delete the link between the `Country` nodes on the left
    and right as you are now looking up this value in the functoid.

12. The map should now look like the following screenshot.

There are some limits in the bridge configuration **user interface (UI)**
that can make changing configuration difficult. It is worth remembering
that the bridge is simply an XML configuration file with associated
configuration files for each source and destination present on the bridge.

For example, there is no way through the UI to change the database details
for an existing lookup added in an Enrich stage. To do this, though, you
simply need to open the `LookupProviderConfigurations.xml`
file and edit the connection details. It is also important to notice that the
username and password details for the connection are actually stored in
this file, and it should therefore be treated with some care.

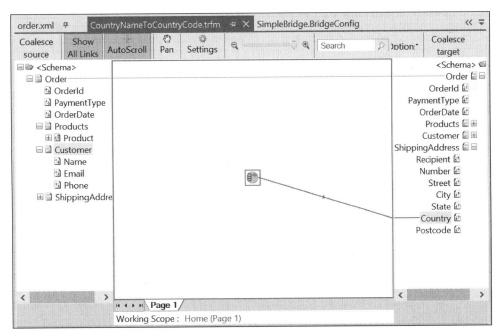

The country-code map

13. The last step needed is to associate the map with the bridge. Double-click on the bridge file to open it (the `MessageFlowItinerary.bcs` file in the Solution Explorer).

14. Double-click on the **SimpleBridge** component.

15. Scroll down to the **Transform** stage and click on the **XMLTransform** box.

16. In the **Properties** window, click on the elipsis (**...**) next to the **Maps** property.

17. The map you just created should be shown in the dialog box; just check the **Selected** column to enable it.

18. Click on **OK**.

# Routing

Now let's look at another common scenario for messaging—routing. Here, a message needs to be delivered to one of a number of potential endpoints depending on some criteria. That criteria could be based on the property of the message (such as where it came from) or a property in the message (a data item such as `country`). Such content-based routing is easily achievable with BizTalk Services, as we'll see.

It should be noted that a message cannot be sent to more than one endpoint. This is something BizTalk Server is capable of, but currently BizTalk Services is not. Instead, BizTalk Services allows you to choose between destinations based on the routing rules you configure.

Look at the design in the *Routing bridge pattern* figure of the *Enrichment* section again. Notice that there are two possible destinations. We will now configure the message flow itinerary to route the message to the `Americas` destination if the `Country` property is `USA`; otherwise, we'll route to `Europe`. To do this, we need to perform the following steps:

1. Click on the arrow connecting the bridge to the `Americas` destination.

2. In the **Properties** window, click on the **Filter Condition** property.

3. Enter `MappedCountry = '844'` or `MappedCountry = '124'`.

4. Now click on the arrow connecting to the `Europe` destination.

5. In the **Properties** window, click on the **Filter Condition** property.

6. Enter `MappedCountry = '826'`.

You can change the order in which the routes are evaluated by clicking on the bridge, and in the **Properties** window, click on the ellipsis (**...**) next to the **Route Ordering Table** property. A dialog box, as in the following screenshot, will be shown:

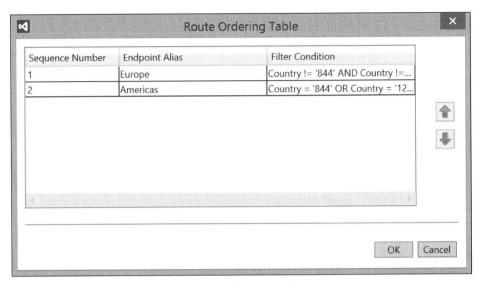

Changing the route order

By using the up and down arrows, the evaluation order can be changed so that you can ensure that the first matching condition you want is where the message will be routed to.

# Trying it out

As the bridge sends messages to one of the two Service Bus queues, you need to create these first in the Azure Management Portal. Create two queues, one called `europe` and the other `americas`. The connection information for these queues then needs to be set on each of the queue destinations on the bridge. The **Runtime Address** property for each takes the following form:

```
sb://<your namespace>.servicebus.windows.net/Europe
```

The Authentication property also needs to be configured. The Token Provider type should be set to Shared Secret and the Issuer Secret set to the ACS Key for your Service Bus namespace.

You're now ready to deploy the solution. Do this in the normal way, and once deployed, you will have an HTTPS endpoint deployed to which you can post messages.

To send a message into the deployed bridge, you can use the BizTalk Service Explorer, which provides a number of useful features for managing and testing your solutions. It is an extension to Visual Studio and can be set up as follows:

1.  Launch Visual Studio 2012, and in the **Tools** menu, select **Extensions and Updates...**.

2.  Click on the **Online** link in the top-left corner, and in the search box, enter `biztalk service explorer` as shown in the following screenshot:

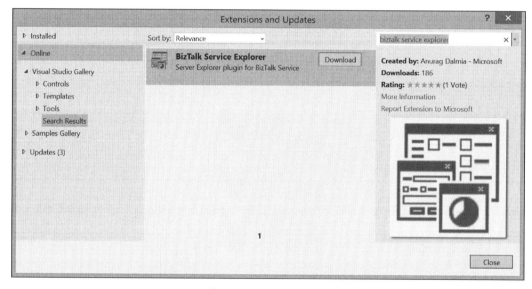

Installing BizTalk Service Explorer

3.  Click on the **Download** button, which will download an MSI file. Double-click on it to install and relaunch Visual Studio.

4.  On the **View** menu, select **Server Explorer**.

5.  The **Server Explorer** window will have a new node, **Windows Azure BizTalk Services**; right-click on it and select **Add BizTalk Service...** as shown in the following screenshot:

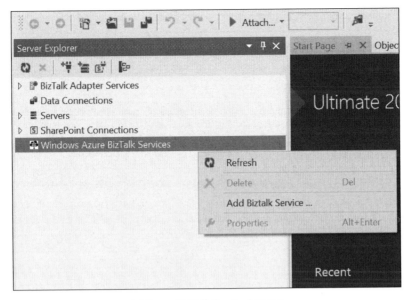

Adding a BizTalk Service instance

6. In the dialog box that appears, enter the details of your service instance as shown in the following screenshot, replacing the details with your values as appropriate, and click on **OK**:

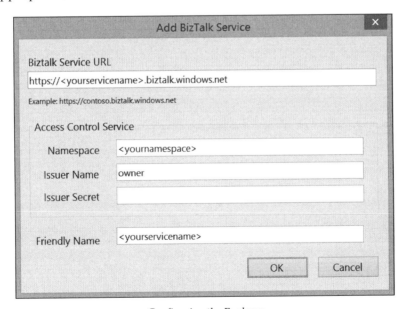

Configuring the Explorer

7. Now that you have the Explorer set up, expand the **Bridges** node and right-click on the bridge you just deployed and click on **Send Test Message…**.

8. In the dialog box that appears, paste the test message from the start of this chapter as shown in the following screenshot and click on the **Send** button:

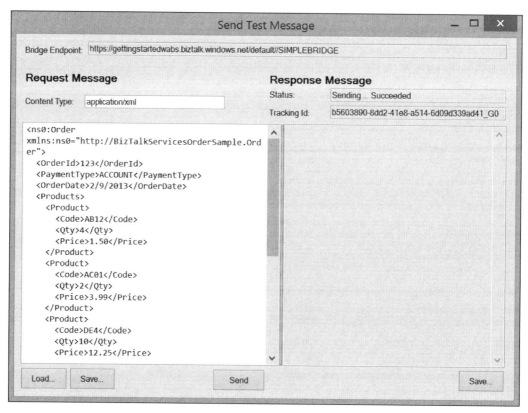

Testing a bridge

When you do this, remember that depending on the value you set Country to in the input message, you can direct messages to either the Europe or the Americas queue, as shown in the *Routing bridge pattern* figure of the *Enrichment* section, by using the values UK, USA, or CANADA.

To view the contents of a queue, you can use the Service Bus Explorer application that you can download from `http://code.msdn.microsoft.com/windowsazure/Service-Bus-Explorer-f2abca5a`. The following screenshot shows the `Europe` destination queue containing the message we just posted into the bridge. Notice the `Country` node in the message contains the value `826` from the SQL lookup table, replacing the value `UK` that was in the original message.

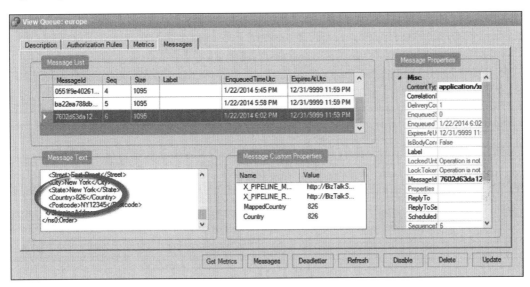

Viewing a message in the queue

Try changing the test message country to one that is not in the lookup table to see what happens. If you enter a country that does not exist, the lookup will fail. You will actually get a 500 HTTP response code back with a SOAP fault error message **Lookup returned no results**. Now try changing the value of `UK` in the lookup table to, say, `123`. What will happen now is that the route will fail as there is no match to either destination. You will get back the same HTTP 500 code, but this time with a SOAP fault of **No Filter matched for the message**. Have fun experimenting on your own!

# Brokered messaging

If you refer back to the table in the *Enrichment* section, you may have been intrigued by the Brokered property type. Messages that flow through bridges received from Service Bus are based on the Service Bus BrokeredMessage type (see `http://msdn.microsoft.com/en-us/library/microsoft.servicebus.messaging.brokeredmessage.aspx` for more details). This class provides a number of properties that are exposed in BizTalk Services such as `CorrelationId`, `MessageId`, and `SessionId`. What is really interesting about this is that when you are using the Service Bus Queue or Topic destination, properties you create in the bridge (any property, not just brokered properties) or properties set on received messages when Service Bus is the source are not just accessible inside the bridge, but outside of it as well. This is very useful for passing state from BizTalk Services to a downstream application consuming messages from a queue as that application will be able to see the properties you set in the bridge.

Note that though this is only applicable for Service Bus sources and destinations, if you were to chain one bridge to another, for example, you would not be able to pass these properties because chaining actually makes calls over HTTP and thus loses the context. Instead, if you wish to pass properties between bridges and from a bridge to another application, you must write the properties to the message header (in the case of HTTP) in order to preserve them. However, if you are chaining via Service Bus, then properties set in the first sending bridge will be accessible in the second receiving bridge.

# Summary

In this chapter, we have taken a closer look at the fundamental construct of Windows Azure BizTalk Services—the bridge. We've seen how bridges can be configured to perform a range of integration activities and how to perform content and context-based routing. While we have looked at most of what bridges can do, in the next chapter, we'll revisit bridges and look at how to perform custom logic on the stages in a bridge using message inspectors. We'll then look at how to track and record the message properties you create (tracking) and how to batch messages together for sending.

# 4
# Enterprise Application Integration

A middleware system or service that enables applications to connect to each other to exchange data is known as Enterprise Application Integration or EAI. In BizTalk Services, EAI is oriented towards a developer persona and Visual Studio is the primary tool for development and deployment of services. Integration between applications is possible using a bridge for messaging. We will explore the e-commerce example from the first chapter in more detail as we look at the concepts.

Specifically, in this chapter, we will focus on the following topics:

- Understanding EAI capabilities in Azure
- Understanding bridges, sources, and destinations
- Understanding custom code using message inspectors
- Understanding hybrid connectivity

## Enterprise application integration scenarios

Consider the following scenarios:

- Contoso is a movie ticketing company and sells tickets through Point-Of-Sale terminals across different cities. They wish to consolidate their end-of-day sales data from the terminals to their SAP Line-of-Business system. In the absence of any form of middleware, the POS data needs to be collected manually and the data has to be merged and converted to a format matching the target system. Using EAI, the entire process can be automated and set up in a matter of minutes with BizTalk Services.

- Fabrikam is a software vendor and uses Salesforce to manage their customer pipeline and sales orders. All approved orders from Salesforce need to be managed centrally in their ERP system like Oracle, which resides on-premises. Using hybrid connectivity all connections to Oracle on-premises can be managed using BizTalk Adapter Services.

- Northwind is an online retailer who manages an e-commerce website for customer purchases. They also receive bulk orders from event firms and corporates for their goods. Northwind needs to develop a solution to validate orders and also route requests to the right inventory location for delivery of the goods. Using EAI in BizTalk Services, they develop a common solution to process purchase orders from consumers over XML as well as purchase orders in EDI from event firms.

Each of these scenarios can be modelled as an EAI solution on BizTalk Services. The incoming requests (ticketing sales, sales orders, and invoices) can be XML or flat file messages and need to be transformed into a target Line of Business format and routed to the on-premises systems. If the destination is on-premises, then relay endpoints are set up using hybrid connectivity.

# EAI in BizTalk Services

Let's look at each of the concepts in more detail and understand their capabilities.

# Sources

Sources receive a message from an external application. BizTalk Services v1 supports five common out-of-the-box sources: **SFTP**, **FTP**, **HTTP**, **Service Bus Queue**, and **Service Bus Subscription**. By default, the bridge exposes the HTTPs endpoint secured by the Access Control service. The various sources of bridges are shown in the following screenshot:

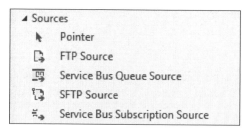

Sources of bridges

# Bridges and the VETER pattern

Bridges are composed of sources, pipelines, and destinations. Pipelines connect two messaging systems and are composed of a series of stages to process the messages flowing from source to destination. The stages perform decoding, validation, enrichment, transformation, and routing of the messages. Each stage can be enabled or disabled for deployment from the Visual Studio properties pane. The set of stages are fixed, and out of the box, BizTalk Services v1 enables the **VETER** pattern. This is shown in the following screenshot:

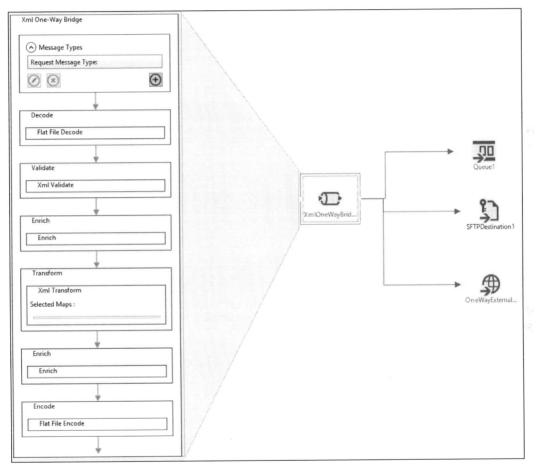

Bridges with VETER pattern

The following are the stages of the VETER pattern:

| Stage | Description |
|-------|-------------|
| Validate (V) | Validate the incoming message against the schema |
| Enrich (E) | Enrich the message with properties promoted from the message header, body, or lookup (see *Chapter 3, Bridges*) |
| Transform (T) | Map the message from one format to another (see *Chapter 2, Messages and Transforms*) |
| Enrich (E) | Enrich the new message post transform |
| Route (R) | Route to one of the target destinations |
| Decode | For flat file processing, decode the message |

# Destinations

**Destinations** are where messages are is submitted to after pipeline processing. In bridges, the route to a destination is based on the SQL-92 expression syntax. A message will be sent to only one destination whose route rule evaluates to true. The route stage is also explained in *Chapter 2, Messages and Transforms*. The various destinations are shown in the following screenshot:

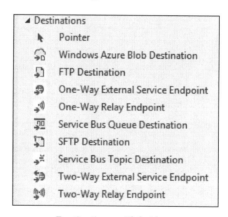

Destinations with bridges

# Attributes of bridges

Here are some interesting attributes of bridges:

- **State**: Pipelines as part of bridges are stateless, that is, at no point during the processing is the message persisted. If there is a crash or restart of the system while messages are inflight, the message would have to be resubmitted for processing. This also means message processing is synchronous.

- **Error handling**: If an error occurs during message processing, a fault is thrown back to the sender of the message to take action. As there is no separate suspend endpoint in EAI bridges, error must be handled at the client side.

- **One-way/Two-way**: Bridges support both one-way and two-way communication. In the case of one-way communication, only the HTTP codes are passed back to the sender of the message. However, in the case of two-way communication, a response message can be sent back. Two-way bridges in BizTalk Services support the VETER pattern on the request side and ETER on the response side. The message on the response side is assumed to be valid since this is coming from the target Line-of-Business system or service. Note that a pass-through-bridge is a special case of a one-way bridge that has only the E-R of the VETER pattern.

- **Message formats**: Messages can be sent in plain old XML, SOAP, and flat file formats. Flat file messages can only be used with one-way bridges. Other formats such as JSON may be added in future, but those different formats need custom code to normalize the data to XML before processing. We will discuss custom code later in this chapter.

- **Chaining**: Multiple bridges can be chained by adding one bridge as the destination of another bridge. This may be used to centralize the processing of messages through a single bridge. For example, multiple bridges may pump messages from different sources all connecting to a single bridge that routes to an on-premises endpoint. Also, selectively disabling stages can enable newer messaging patterns. For example, ETEVR can be achieved by chaining two VETER bridges.

# Hybrid connectivity

Organizations that have made IT investments in ERPs and services on premises may not transition all of their IT assets to the cloud. There is a need to connect to those services and resources using hybrid connectivity from the cloud.

## The BizTalk Adapter Service

The **BizTalk Adapter Service (BAS)** is a service which enables an application running on-premises to receive data from the cloud. The on-premises applications such as ERPs and BizTalk Server can be exposed outside of the corporate network using Service Bus relays for hybrid connectivity. The Service Bus Relay service on Azure acts as an intermediary where a client and an on-premises service can connect with each other. The client in this case is the BizTalk Services' bridge and the service running on-premises is the BizTalk Adapter Service talking to an ERP. Once the BizTalk Adapter Service and the bridge authenticate with the Service Bus Relay service, all messages from the bridge are forwarded to the BizTalk Adapter Service.

# The BAS architecture

The overall BAS architecture is shown in the following figure:

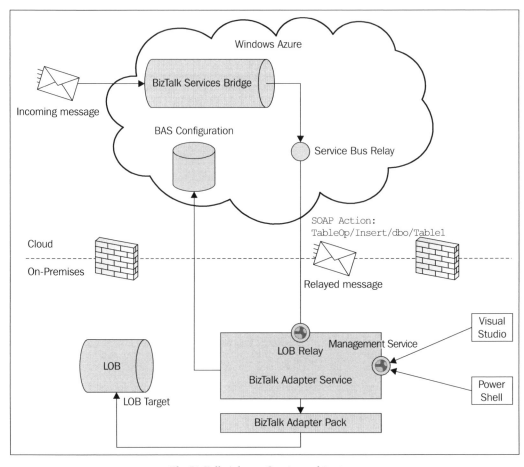

The BizTalk Adapter Service architecture

The BizTalk Adapter Service runs on the client machine and is hosted in IIS to handle management operations such as start/stop of endpoints as well as runtime operations to route messages from the cloud to the on-premises systems. There is one management service along with one or more runtime services that can be managed using the management service. The meta-data of the relay configurations is stored in the storage account of the BizTalk Services deployment specified during installation. The BizTalk Adapter Service relies on the BizTalk Adapter pack to connect to Line-of-Business (LOB) systems such as Oracle DB, Oracle EBS, SAP, Siebel, and SQL Server.

Management operations are exposed through Visual Studio Server Explorer or through PowerShell cmdlets, both of which talk to the **BAService**, the application hosting the `ManagementService.svc` service in IIS on the on-premises machine where BizTalk Adapter Service is installed.

Runtime operations are managed as per the Service Bus Relay by creating new applications in IIS hosting `RuntimeService.svc`. Each LOB can create a new relay or use an existing relay configured for another LOB. When there are more than one LOB per relay, the sub-paths in the runtime address beyond the relay URL help direct the calls to the right adapter.

When the LOB relay is created, based on user configuration a new or existing application is used in IIS as the service host. The generated WSDL contains the message to be relayed as well as the operation action. Operations such as `INSERT`, `UPDATE`, `DELETE`, and `SELECT` are passed as part of the `SOAPAction` header in the following format: `TableOp/{Operation}/schema/Tablename`. The exact SOAP Action string can be determined from the relay configuration's properties window in Visual Studio.

Every message passing via the relay needs to authenticate with the LOB. Authentication credentials are passed using one of the following four ways:

- Username and password preconfigured and stored in the BAS store
- Active Directory domain credentials
- SOAP header containing credentials of the LOB
- WS-Security credentials

# BAS installation and configuration

BAS installation is part of the BizTalk Services SDK setup. During setup, the URL of the BizTalk Services deployment needs to be entered. This is added to `web.config` under `C:\Program Files\Microsoft BizTalk Adapter Service\BAService`. In Visual Studio Server Explorer, the on-premises management service URL along with the ACS credentials of the deployment need to be entered. Hybrid connectivity can now be set up for each of the following LOBs using a wizard-driven interface as shown:

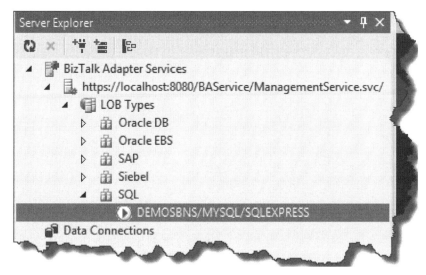

BizTalk Adapter Service configuration in Visual Studio Server Explorer

For example, setting up the relay connectivity with on-premises SQL Server Express running on localhost with DemoDB as the database involves the following steps:

1. In the **Server Explorer** BAS view, right-click on **LOB Types | SQL** and choose **Add SQL Target**.

2. In the pop-up wizard, read the instructions and click on **Next**.

3. Enter values for the server name, instance, and catalog (say `localhost`, `SQLExpress`, and `DemoDB`, respectively). Use Windows authentication or the username and password as configured and click on **Next**.

4. Navigate to **Tables** (or **Views**) is exposed via relay, choose the table, and select **Insert** as the operation. Click on **Properties** to see the WSDL generated. Click on **Next**.

5. Configure the **Runtime security type** when the message passes via relay. These are the four options we mentioned earlier. Enter the credentials and click on **Next**.

6. In **Specify the LOB Relay URL**, choose **Create a new LOB relay** and enter the Service Bus credentials. Enter any name for **LOB relay path** and **LOB relay subpath**. Click on **Next**.

7. Click on **Create** to complete the creation of the relay and the BAS endpoint.

Once the relay has been successfully set up, each connection will appear under the corresponding LOB Type.

# Consuming BAS with bridges

Create a new BizTalk Services project or open an existing one. There are three parts to using the BAS configuration with a bridge:

1. From the Server Explorer, we can now right-click on the relay connection configured earlier and choose **Add Schemas to Project**. The relevant schemas to send and receive the target LOB are added in the project. These can be used for mapping and validation purposes within the bridge.

2. Drag-and-drop the connection from the Server Explorer into the Bridge design surface. This will create the necessary icon to add a destination connection from the bridge.

3. Click on the relay connection and navigate to the **Operations** field in Visual Studio properties window. Expand the view and note down the values to the right of each of the index [0], [1]. Each of these are the SOAP Action values. To add this value, go the Route Action in VS properties and launch the **Route Actions** window. Click on **Add** and in the **Edit Route Action** pop up, enter Expression as the soap action value copied earlier. Choose **Type** as **Soap** and **Identifier** as **Action**. Click on **OK** to accept the changes.

# Custom code in EAI

Now that we understand hybrid connectivity, let's look at one more functionality of bridges, which is to support custom code. Not all capabilities will be available out of the box from BizTalk Services. Customization enables developers to plug in new functionality that augments the existing message flow. For example, we can choose to convert an incoming invoice XML to a user-readable PDF format as well as archive the same for legal reasons.

Customization in a bridge is possible at the stage level, route configuration, or in transforms. Transforms and its customization were covered in *Chapter 2, Messages and Transforms*. In this section, we will look at bridge customization.

# Message inspectors

Message inspectors are custom code hooks for every entry or exit of a stage in a bridge. Custom code must implement the `IMessageInspector` interface:

```
public interface IMessageInspector
{
Task Execute(IMessage message, IMessageInspectorContext context);
}
```

Message inspectors are implemented using the Task programming model in the .NET4 Task Parallel Library. Traces in custom code can be emitted using the `ITracer` interface in the `IMessageInspectorContext` interface.

```
public interface ITracer
{
void TraceEvent(TraceEventType eventType, string format,
  params object[] args);
}
```

The following are key points to remember when developing custom code:

- User code is expected to be resilient, but in some cases, it could throw an exception. In this case, this is treated as a stage level failure and the corresponding track record is generated.
- User code in the VS project must have references to `Microsoft.BizTalk.Services.dll` from `C:\Program Files (x86)\Microsoft Visual Studio 11.0\Common7\ide\Extensions`.
- User code assemblies must be added as reference in the BizTalk Services project with `Copy Local` set to true.
- Whenever the user code assemblies are deployed in BizTalk Services, the service must be restarted as the DLLs need to be reloaded in the .NET AppDomain. The restart option is available in the VS deploy and in the PS cmdlet.
- Artifacts of the bridge such as the schema or map are not accessible within the user code.
- Properties can be defined and promoted in the custom code. In code, they must be string property with C# attribute `PipelinePropertyAttribute` set with the `Name` attribute. This attribute is set in the VS Property Configuration in the Message Inspector configuration window as seen in the *Custom code configuration with bridges* figure in the *Configuring the bridge* section.

# Tracking

Tracking helps in storing interesting properties of a message in the tracking store. The tracking store is an Azure SQL database configured during BizTalk Services provisioning time. All message properties are stored in the `PipelineTrackRecords` and `SourceTrackRecords` tables. Tracking for troubleshooting is detailed in *Chapter 7, Tracking and Troubleshooting*.

To enable tracking at the EAI bridge, select the bridge in VS and choose **Track Properties** from the properties window. The tracked properties can be seen in the BizTalk Services portal's Tracking view.

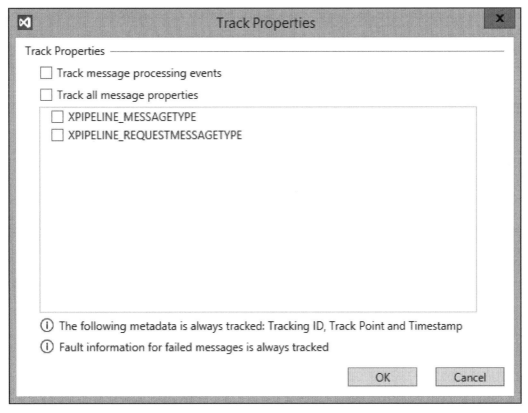

Configure tracking properties with bridges

# Scenario walk-through

Let's revisit the EAI scenario with the following changes.

Northwind is an online retailer who manages an e-commerce website for customer purchases. Instead of processing orders, let's say they now receive invoices from their suppliers for the goods sold. For readability and regulatory reasons, they need to store this in PDF format in an on-premises system.

> It is assumed that the BizTalk Services SDK has been installed and Visual Studio shows the following projects:
> - BizTalk Service project: Create/Deploy bridges, schemas, and maps
> - BizTalk Service artifacts project: Create/Deploy schemas, and maps

# Prerequisites

Northwind creates a new BizTalk Services deployment. See *Chapter 1, Hello BizTalk Services* on creating a BizTalk Services deployment and registering the BizTalk portal. We are going to use the PDFTemplate utility from `pdftemplate.codeplex.com` to generate PDF-formatted invoice messages. The utility is available under the GPLv2 license.

# Solution

The solution would take the invoice XML and generate the PDF in a blob store. To get started, let's first create the schema, add the code to generate the PDF, and finally plug that logic into the bridge configuration.

## Creating a schema

Create a sample schema for use with the incoming message. The sample used in this flow is provided along with this chapter. Perform the following steps:

1. Using the Visual Studio schema editor, create a simple schema called `InvoiceSchema.xsd` for the invoice.

2. From the Visual Studio command prompt, run the following command to generate `InvoiceSchema.cs` for this xsd:

   ```
   xsd InvoiceSchema.xsd /classes
   ```

3. We will load this `InvoiceSchema.cs` file in the next step.

# Creating custom code

Let's now add the custom code to generate the PDF for the invoice we just created:

1.  Create a C# class library project and add a reference to `Microsoft.BizTalk.Services.dll` and the PDF dependencies.

2.  Implement a class for the `IMessageInspector` interface. In the example, we have extracted the message body to the `Order` message object as follows:

    ```
    string bodystr = GetMessageBody(message);
    Order order = DeserializeMessageToOrder(bodystr);
    ```

3.  PDFGenerator describes the layout of the PDF structure in `layout.xml` as required by the codeplex tool. The XML itself is passed within the custom code DLL as an embedded resource and must be extracted before use as shown in the following code snippet:

    ```
    Stream stream = System.Reflection.Assembly.GetExecutingAssembly().
    GetManifestResourceStream("layout.xml");
    byte[] bytes = new byte[(int)stream.Length];
    stream.Read(bytes, 0, bytes.Length);
    File.WriteAllBytes(System.IO.Path.GetTempPath(), bytes);
    ```

4.  Fill in the header, body, loop, and footer data of the PDF file using the data from the `Order` object. Use the PDFGenerator using the information shown in the following code snippet:

    ```
    PDFTemplateItextSharp pdfgen = new PDFTemplateItextSharp(xmlToPdfT
    emplateFilePath);
    pdfgen.Draw(GetHeaderData(order), GetLoopData(order),
    GetBodyData(), GetFooterData());
    ```

5.  Write the PDF data back as a new message shown in the following code snippet:

    ```
    message.Data = new System.IO.MemoryStream(pdfdata);
    message.ContentType =
      new System.Net.Mime.ContentType("text/plain");
    ```

6.  Sign the assembly output from the project. Note that all the DLL dependencies are needed to be signed.

# Configuring the bridge

Perform the following steps to configure the bridge:

1. Add a BizTalk Services project to the same solution and add references to the signed custom code.

2. Add a bridge, in this case, a **Pass-Through** bridge, and call it Invoice2PDF. Also add the route for the message to a Windows Azure blob destination. Let's call the blob storage **PDFArchiveBlobs** as shown in the following figure:

Invoice2PDF bridge sample in BizTalk Services project

3. In the bridge properties, fill the required fields. For the bridge, add the BizTalk Services **Runtime Address** and **Routing Table** values. For the blob, add the **Shared Access Signature URL**. Click on the bridge, and from the properties window, open **Track Properties** on the bridge window to enable tracking.

4. Get the full qualified assembly name of this custom code. If this is specified incorrectly, you will get an error during deployment. You can use the **GetAssemblyQualifiedTypeName** sample in MSDN Code Gallery, `http://code.msdn.microsoft.com/windowsazure/Windows-Azure-BizTalk-EAI-56915d1c/view/SourceCode`, or alternatively, run `sn -T` on the DLL to determine the public key and determine the full qualified name. It should look something like this:

```
PDFGenerator.PDFGeneratorUtil, PDFGenerator,
  Version=1.0.0.0, Culture=neutral,
  PublicKeyToken=xxxxxxxxxx
```

5. Double-click on the pass-through bridge and click on the Enrich stage. From the properties window, click on **On-Exit Message Inspector** and add the full qualified name in the **Specify Custom Code Inspector** pop-up window, as shown in the following screenshot:

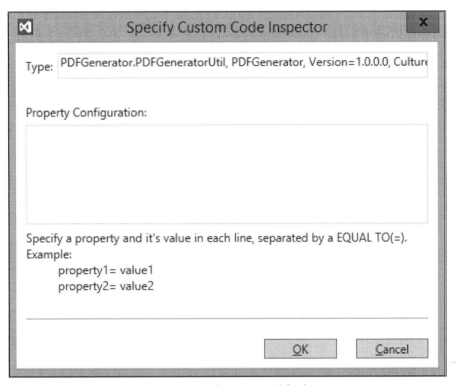

Custom code configuration with bridges

# Deploying the bridge

We can now deploy the bridge to the BizTalk Services deployment with the following steps:

1. Launch VS and select the **Deploy** command from the **Build** menu and enter the BizTalk Service deployment details.

2. If the project is being deployed more than once, you need to check the **Refresh Service after deploy** checkbox so that the updated custom code DLLs are picked-up.

The VS output block will be similar to the one shown as follows:

VS output for bridge deployment

## Sending messages

Use the Message Sender tool from MSDN Code Gallery samples for BizTalk Services to send messages to the bridge, or you can also download the BizTalk Services Explorer plugin for VS Server Explorer. This allows you to explore the deployment and also send test messages.

Once the messages are sent successfully, go to the container with the SAS URL and save the blob locally. Rename the file to a .PDF extension and you should be able to view the archived PDFs.

## Viewing tracking data

Click on the Tracking view in the navigation bar of the BizTalk Services portal to see the status of message flow on the bridge. It's also possible to view this information from the BizTalk Service Explorer in VS on a per-bridge basis.

# Summary

In this chapter, we started with the basic concepts of EAI on Azure, notably bridges, sources, destinations, hybrid connectivity, and custom code. We walked through a simple scenario in BizTalk Services generating PDF invoices and archiving in a blob store. It is possible to encounter an error while using custom code in bridges. We will cover aspects of troubleshooting in *Chapter 7, Tracking and Troubleshooting*.

In the next chapter, we will look at another key scenario supported in BizTalk Services — integrating across businesses using B2B capabilities.

# 5
# Business-to-business Integration

A transaction between two companies in a buyer-seller relationship is business-to-business (B2B) as opposed to a B2C relationship between a company and its consumer. In the context of integration, B2B is about two organizations agreeing on a set of well-defined transactions for the exchange of business information. The B2B process starts off with a business negotiation or an agreement, which further translates into the technical details at the message flow level.

In this chapter, we will focus on the following topics:

- Understanding B2B integration in the context of Azure
- Understanding the capabilities of the service with partner, agreement, batching, and tracking
- Walkthrough of a real-world B2B scenario using BizTalk Services on Azure

## Basic concepts of B2B

Consider a retailer, Contoso, who wishes to procure stock from a supplier, Northwind. From a technology standpoint, the message flow between Contoso and Northwind can be explained using the following flow:

- Contoso and Northwind agree to enter into a business relationship (usually over phone/e-mail/meetings). The legal teams draft the memorandum of understanding (MoU) between the partners.
- The partners exchange information on how one would receive/send payments, including cheque and bank details.

- Both partners enter a setup phase where each of their IT departments configures the B2B system to enable the exchange of electronic data interchange (EDI).

- Both partners exchange sample test messages to validate the configuration, and after sufficient testing, they agree to move ahead with production.

- In production, the following steps are carried out:

  ○ Contoso raises a purchase order from their Dynamics AX ERP system and sends it to Northwind

  ○ Northwind receives the order and acknowledges that they have received the order

  ○ Northwind processes the order and looks up the inventory in their SAP ERP to determine whether they can service the request

  ○ If the order can be serviced, Northwind send the goods to a third-party logistics company (3PL) who handles the shipment of goods

  ○ Northwind replies with the shipment details to Contoso

  ○ Contoso can optionally acknowledge the receipt of the shipment notice

  ○ Northwind sends an invoice to Contoso for the cost of goods sold

- Contoso sends payment to Northwind based on the financial terms agreed upon.

The preceding flow can be visually represented as follows:

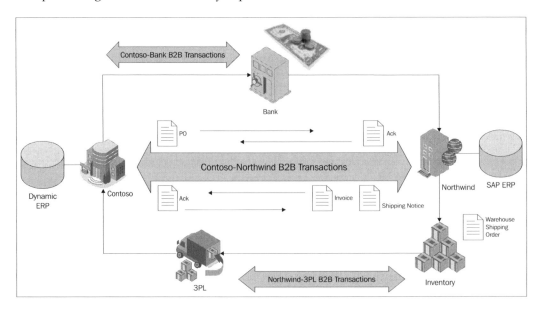

Each of the arrows in the preceding diagram represents a set of messages exchanged between Contoso and Northwind. Each message type is identified by a document name or transaction set (for example, PO/X12 850) and has a defined format based on the protocol used. The protocol, in addition to the structure of the message, governs the set of messages exchanged back and forth. In the preceding example, we can write the first transaction as "Contoso raises an EDI X12 purchase order (850) to Northwind". Here, X12 is the EDI format and 850 is the purchase order message type.

# Common interaction models

There are two common ways to integrate trading partners for B2B. They are listed as follows.

# Direct enterprise integration

In this model, both trading partner organizations have in-house IT and can directly transact EDI messages. There are systems in place to send and receive EDI transactions over point-to-point protocols without any mediator or middleman.

# Service provider integration

In this model, one of the trading partners is a small-to-medium business player who cannot afford in-house IT. In order to facilitate EDI interactions, there is a middleman, or the EDI service provider, who acts as a liaison between the two partners. The service provider talks about EDI to a trading partner on one end and transacts non-EDI (such as XLS/XML) with the other trading partner. The service provider charges a fee based on transaction size/volume or the complexity of the protocols used.

**Value Added Networks (VANs)** are specialized networks offering end-to-end B2B services in a service provider integration. VANs manage the hosting of servers and software to process EDI traffic and usually charge based on message volume. They could specialize solutions for a particular vertical or offer generic message processing services.

# Industry standards and protocols

Various organizations define protocols for different industry verticals. Each protocol governs the set of messages exchanged, the acknowledgements transferred, and the error behavior between trading partners.

The following are some of the most common protocols and their standards organization supported in BizTalk today:

| Standards body | Website | Protocol | Industries |
|---|---|---|---|
| ANSI ASC X12 | `http://www.x12.org/` | X12 | In the US: manufacturing, retail, government, and transportation |
| | | HIPAA | Healthcare, insurance |
| UN/ CEFACT | `http://www.unece.org/cefact/edifact`<br>`http://www.gefeg.com/jswg/` | UN/ EDIFACT | In Europe: manufacturing, retail, government, and transportation |
| SWIFT | `http://www.swift.com` | SWIFT | Financial transactions for treasury, trade, and banking |
| RosettaNet | `http://www.rosettanet.org/` | RosettaNet | Supply chain |
| OAGi | `http://www.oagi.org` | CIDX | Horizontal framework for several verticals |
| PIDX | `http://www.pidx.org/` | PIDX | Oil and Natural Gas |
| HL7 | `http://www.hl7.org/` | HL7 | Healthcare |

X12 is heavily used in the US, while EDIFACT is more popular in Europe and Asian countries. Both these protocols are supported in BizTalk Services today. The remaining protocols are available in BizTalk Server.

# Concepts in BizTalk Services B2B

B2B in BizTalk Services is all about processing EDI and non-EDI messages between trading partners. It is meant to make B2B integration simple, powerful, and flexible using Azure. B2B, by its nature of protocols, formats, and transport, tends to be complex in configuration; with Azure, service configuration of an agreement between partners is simple. It's easy to extend the service and connect with other technologies such as SharePoint and mobile services to build a rich and powerful solution.

The top-level concepts in B2B/EDI include the following:

- EDI message structure
- Partners and agreements
- Property promotion in EDI

- Batching
- Tracking and archiving
- Extensibility and object model API

# EDI message structure

EDI messages (either X12 or EDIFACT) have a nested structure, compartmentalizing transactions for ease of understanding by the receiver. Every structure has a header and trailer to identify the start and end of the nested structure. The following outlines the nesting structure:

- **Interchange**: This is the outermost envelope with a header and trailer. It identifies the sender and receiver of the message as well as the date/time when the message was sent. In the case of X12, the ISA and IEA are the header and trailer respectively; in EDIFACT, the UNB and UNZ form the header and trailer.

- **Group**: A group segment in an interchange is a set of transactions clubbed together by their function. A group starts with a header (GS in X12 and UNG in EDIFACT) and ends with a trailer (GE in X12 and UNE in EDIFACT). Unlike X12, groups are optional in EDIFACT; but when groups are present, they must contain all transactions of the same type (for example, all purchase orders).

- **Transaction set**: A transaction set in a group is the message of a given type (for example, a purchase order message) with segments detailing transactions such as item quantity or price. A transaction set starts with a header (ST in X12 and UNH in EDIFACT) and ends with a trailer (SE in X12 and UNT in EDIFACT).

The following table illustrates the X12 and EDIFACT headers and trailers:

| X12 header and trailer | EDIFACT header and trailer |
| --- | --- |
| ISA Interchange Control Header | UNA Optional Advice |
| GS Functional Group Header | UNB Interchange Control Header |
| ST Transaction Set Header | UNG Functional Group Header |
| SE Transaction Set Trailer | UNH Message Header |
| GE Functional Group Trailer | UNT Message Trailer |
| IEA Interchange Control Trailer | UNE Functional Group Trailer |
| | UNZ Interchange Control Trailer |

# Partners and agreements

The following are the key concepts used with agreements:

- **Partner**: A partner is an organization with which a trading relationship is established. Each partner has one or more business units referred to as **business profiles**. Every business profile has an identifier (for example, a DUNS ID or phone number) that is unique and added to each message exchanged between the partners.

- **Agreement**: An agreement represents the technical settings of message exchange between partners. An agreement is established between two business profiles of partners. The agreement is composed of send settings (send of Contoso, receive of Northwind) and receive settings (receive of Contoso and send of Northwind) in the trading partner relationship. An agreement refers to a protocol such as AS2, X12, EDIFACT, and so on, based on messaging requirements. An agreement also identifies the schemas and requirements around tracking and batching. Deployment of the agreement results in two bridge deployments and, therefore, endpoints in Azure that can receive EDI messages and send XML and vice versa. The agreement definition in the BizTalk Portal has the following:

  - **Transport**: This is the send side or the receive side transport of the bridges.
  - **Protocol**: This is the EDI protocol under use in the bridge.
  - **Transform**: These are the maps used while a message is processed in the bridge.
  - **Route**: This is the target destination where the bridge will route the message to the recipient of the message.
  - **Inbound URI**: This is the address of the send side bridge which receives messages to be sent to the destination partner.
  - **Suspend endpoint**: This is the endpoint which would process the message if there are errors in the processing of the EDI message. This endpoint can be used to build repair-resubmit scenarios.

- **Partnership**: At least one agreement between partners constitutes a partnership between the partners. The concept of partnership is exposed only through the trading partner management API.

- **Agreement template**: An agreement template is a unit of re-use where commonly repeated settings can be captured as a template and applied while defining an agreement for rapid configuration. An agreement template is associated with a profile. Each template definition identifies the hosted partner in the definition to determine the direction of settings to be applied while creating an agreement.

- **AS2 agreement**: AS2 agreement refers to the AS2 transport settings agreed between two partners adhering to the RFC 4130 standard. In a nutshell, AS2 allows messages and acknowledgements to be transmitted in compressed, signed, or encrypted form over HTTP or HTTPS. Signing and encryption is supported using certificates. AS2 is not specific to the payload and can be used for both B2B flat files and EAI XML messages.

> If the endpoints of one partner, say Partner A, require HTTPS for AS2 traffic, then Partner A's deployment certificate needs to be added to the "Trusted People" certificate store of Partner B. Alternately, if the AS2 messaging from Partner A is to an HTTPS endpoint of Partner B, then Partner B's public certificate needs to be added to the certificate store of Partner A's deployment. Both these certificates need to be uploaded to the BizTalk Services deployment certificate store using `PSCmdlet` loaded from BizTalk Services Tools. Here is the sample code to add the certificate using PowerShell:
>
> ```
> PS C:\>Import-Module 'C:\Program Files\Windows Azure
> BizTalk Services Tools\Microsoft.BizTalk.Services.
> PowerShell.dll'
> PS C:\>Add-AzureBizTalkArtifactCertificate -
> AcsNamespace myAcs -IssuerName owner -IssuerKey
> 193194218484a= -FilePath D:\sample.cer -ArtifactPath /
> sample.cer -certificateStore TrustedPeople
> ```
>
> If self-signed certificates are generated by using `makecert.exe`, ensure that you pass the `-pe` and `-key` exchange parameters to the command to make the certificate exportable and usable for encryption purposes.

- **X12 and EDIFACT agreement**: Both X12 and EDIFACT are supported on BizTalk Services. The choice of the agreement is available through a combobox selection during agreement creation. For each application protocol, the agreement settings consist of a series of schema selection, acknowledgement configuration, control number configuration, batching, character sets and separators configuration, and configuration of message validation.

# Property promotion

Certain properties in the EDI envelopes we discussed earlier are autopromoted and are available for use in routing and tracking scenarios. They can also be used to determine the agreement endpoints programmatically.

Listed in the following tables are the promoted properties in EDI and AS2:

| Property | Location | Description |
| --- | --- | --- |
| Message Type | X12 Receive | Numeric value identifies type of message, for example, 850 PO, 810 Invoice |
| AgreementName | X12 Receive | Name of the agreement |
| ISA 5-8 | X12 Receive | X12 ISA envelopes |
| ISA 9, 10, 12, 15 | | |
| GS01-08 | X12 Receive | X12 GS envelopes |
| ST01 | X12 Receive | X12 transaction set message type |
| ST03 | X12 Receive | X12 transaction set version |
| AS2-To, AS2-Version, Mime-Version, AS2-From | AS2 Send/Receive | AS2 header properties |
| Content-ID, Content-Type, Content-Transfer-Encoding | | |
| Disposition-Notification-To, Disposition-Notification-Options | | |
| Content-Description, Content-Disposition, Receipt-Delivery-Option | | |
| SystemRequestID | X12 Send, X12 Receive, AS2 Receive | ID to track the message flow in the bridge |
| MessageReceivedTime | X12 Send, X12 Receive, AS2 Receive | Date and time of incoming message |
| SourceType | X12 Send, X12 Receive, AS2 Receive | FTP, HTTP, or AS2 per configuration |
| SourceName | X12 Send, X12 Receive, AS2 Receive | The name configured for source; this is an autogenerated name in EDI agreements |
| AgreementID | X12 Send, X12 Receive, AS2 Receive | Agreement ID as displayed in the agreements list view in the BizTalk portal |
| UNA_Segment | EDIFACT Receive | UNA Segment with characters as separators and indicators |

| Property | Location | Description |
|---|---|---|
| UNB_Segment | EDIFACT Receive | Interchange header segment |
| UNB2_1 | EDIFACT Receive | Sender Identification |
| UNB2_2 | EDIFACT Receive | Sender Code qualifier |
| UNB2_3 | EDIFACT Receive | Sender Reverse Routing address |
| UNB3_1 | EDIFACT Receive | Receiver Identification |
| UNB3_2 | EDIFACT Receive | Receiver Code qualifier |
| UNB11 | EDIFACT Receive | Test indicator |
| UNG_Segment | EDIFACT Receive | Functional group header segment |
| UNG1 | EDIFACT Receive | Functional group identifier |
| UNG2_1 | EDIFACT Receive | Group Application Sender Identifier |
| UNG3_1 | EDIFACT Receive | Group Application Receiver Identifier |
| UNH2_1 | EDIFACT Receive | Message type identifier |
| UNH2_2 | EDIFACT Receive | Message type version number |
| UNH2_3 | EDIFACT Receive | Message type release number |

 AS2 Send is sending the message to the partner; hence, the properties are not usable directly. In the case of X12 Send, we can route the message to a blob store or Service Bus queue and base our action on the properties promoted.

# Batching

Batching is a concept where messages are accumulated based on selection criteria and released once the desired event known as the release criteria is met. Customers use batching of messages to aggregate messages for reasons of cost, compatibility, and convenience. In the early days of VAN, messages were charged on size and customers were optimized by sending batches of messages. An example of this is airlines charging the same for all containers, whether they contain 2000 kg or 5000 kg. It makes sense for freight forwarders to batch requests from multiple shipments into a single container to save costs. Few mainframe systems of trading partners cannot accept messages more than 250 KB. It is important to break messages such as these to sizes less than or equal to 250 KB before sending them to the target system.

As part of BizTalk Services B2B, customers can configure and manage a batch in an agreement's send side configuration (debatching is already part of the system as the interchange format is implicitly known). An agreement can contain zero or more batch definitions. Every batch has a state—it can be enabled, disabled, or in error. Batches are enabled when the corresponding agreement is deployed or the start command is used explicitly. When batches are stopped, messages in the batch are flushed out. Each batch definition contains the selection and the release criteria.

## Selection criteria

The selection criteria are used to select a message for batching. In BizTalk Services, customers can promote properties and use them in expressions to select the message to be added to one or more batches. If the property of an incoming batch matches more than one definition, then the message is copied to these many batches.

## Release criteria

The release criteria determine when the batch should be released. One of the following can be used as release criteria:

| Release criteria | Description |
| --- | --- |
| Size | The maximum size of the message (excluding interchange and groups) in bytes with UTF-8 encoding |
| Count | The number of transaction sets in a batched message |
| Schedule | The configuration of the occurrence and recurrence values to release messages periodically |
| Timeout | Inter-message idle timeout when a batch needs to be released |
| Interchange size and schedule | Either size or based on schedule |
| Interchange size and timeout | Either size or configured timeout |
| Message count and schedule | Either count or based on schedule |
| Message count and timeout | Either count or configured timeout |
| Interchange size and message count | Either size or count |

Messages from a batch are released when either the release criteria are met or the batch has been stopped from the BizTalk portal. If the release criteria are met but no messages are in the batch, a null message is not sent to the send endpoint. If the send endpoint is not available when the batch releases a message, the message is passed to the suspend endpoint. In the worst case, if the suspend endpoint is also down in spite of retries, the set of unbatched transaction sets are held in the system and released to the suspend endpoint the next time it becomes available. Also, note that for an agreement to be deleted, the batches defined in the agreement must not have any messages.

# Tracking and archiving

Tracking helps in storing interesting properties in a message and archiving messages in stores. Both tracking and archiving are settings enabled as part of an agreement configuration. In the case of B2B, the system and user-promoted properties are tracked and written to the Azure SQL Tracking database. In addition to messages, the system tracks properties with which a message can be correlated with its acknowledgement. This would let users know if an X12 has received a technical or a functional acknowledgement and if an AS2 message has received its MDN NACK or ACK message. Tracking identifies the list of batches that are active with the list of messages currently held in the batch. It also identifies the history of batches that were dispatched on the send side.

Archiving is supported in the following cases for X12 and AS2 as illustrated in the following diagram:

- Just before sending a message
- Immediately after receiving a message

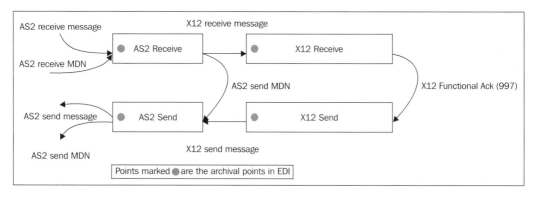

Archiving is configured in the **General Settings** section in the agreement. The messages can be accessed from the BizTalk portal's tracking view option by clicking on the relevant message entry and choosing **Details**. From the message info pop-up view, you can select the entity to download and click on the **Download** option.

# Non-repudiation

**Non-repudiation of Receipt** (**NRR**) is supported in AS2 using both tracking and archiving. NRR is required in dispute settlement scenarios where, for example, a supplier may not process a PO or claim not to have received the order or the payment. NRR ensures that the AS2 message is stored and the incoming Message Integrity Check (MIC) value or hashcode is validated against the MIC of the stored AS2 message. This validates that the other partner has indeed received and processed the message. If the NRR option is turned on and if the tracking or archiving fails, then the message processing also fails (unlike in cases without NRR, where message processing can continue even if there are tracking errors). To enable NRR, check the **Enable NRR** option in the **General Settings** page of the AS2 agreement, as shown in the following screenshot:

# Extensibility

Extensibility in B2B is possible by using the public API for **Trading Partner Management Object Model** (**TPM OM**). We will cover TPM OM as part of the overall extensibility in BizTalk Services using API.

# Scenario walk-through

Let us revisit the scenario we began with in this chapter; we will add a service provider to illustrate the concepts. A service provider, as we know, is an expert in EDI and will act as a middleman to help suppliers connect to large retailers. In this example, Contoso, a large retailer, wishes to procure stock from supplier Northwind. Since Northwind is not aware of EDI, they approach a service provider, Fabrikam, to help them connect with the retailer Contoso.

## Ecosystem players

There are three players in our example. Contoso is a large retailer, Northwind is the supplier connecting to Contoso, and finally, Fabrikam is the EDI service provider providing EDI services to Northwind.

## Provisioning BizTalk Services

Fabrikam creates a new BizTalk Services deployment. See *Chapter 1*, *Hello BizTalk Services*, on creating BizTalk Services deployment and registering the BizTalk Services portal.

Fabrikam adds Northwind's admin e-mail ID as a registered user in the BizTalk Services portal Settings page. This allows Northwind users to log in to the same view as that of Fabrikam.

## Configuring partners – Fabrikam, Northwind, and Contoso

We need to configure the trading partners and the agreement in BizTalk Services. The following steps need to be carried out by Fabrikam to add partners:

1. Once signed in, click on the **PARTNERS** page and then the **Add** button.
2. Enter the **Partner name** as Northwind.
3. Enter **First name**, **Last Name**, **Email ID**, and **Phone** details, if required.
4. Click on **Save**.
5. Repeat the above steps for Contoso and Fabrikam as partner names.
6. Click on each partner, navigate to the default profile, and upload the certificate for AS2 signing and encryption.

The creation of a new partner is shown in the following screenshot:

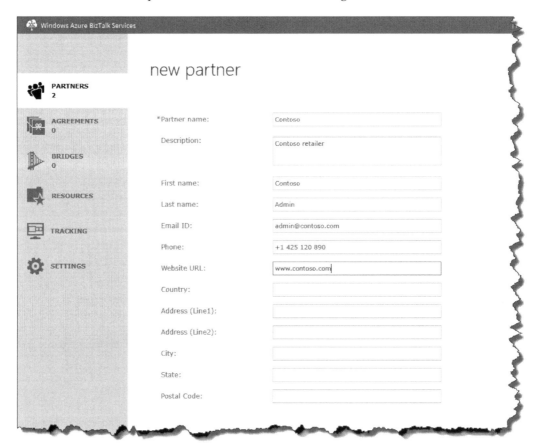

# Configuring the AS2 agreement between Fabrikam and Contoso

Once the partners are created, we need to add agreements. The following steps need to be carried out to add the AS2 agreement between Fabrikam and Contoso:

1. Click on **AGREEMENTS** in the left navigation bar.

2. Click on the **AS2** tab.

3. Click on **Add**.

4. In the **New Agreement** (AS2) page, fill the agreement name as Fabrikam-Contoso agreement and add a description.

5.  Select **Hosted Partner** as Fabrikam and **Guest Partner** as Contoso . Here, **Hosted Partner** is the partner who owns the bridge in BizTalk Services. In this case, Fabrikam owns the BizTalk Services deployment, and hence is the host partner.

6.  Enter **AS2 Identity** of Fabrikam as fabrikam and for Contoso as contoso.

7.  Enable tracking and archiving; the latter is available for Premium SKUs.

8.  Click on **Continue**.

9.  The page is now renamed to Fabrikam-Contoso agreement and you are on the **Receive Settings** page.

10. Following are the steps to configure **Receive Settings**:

    °  Under **Inbound URL**, set **URL Suffix** to endpoint1 and note down the complete URL. This is where Fabrikam would receive messages from Contoso.

    °  Under **Protocol**, open **Message** and choose **Messages should be signed** or **Messages should be encrypted** if applicable. Note that both these options require you to add a certificate in the profiles pages of Contoso and Fabrikam for this purpose. Also, set **Acknowledgement** and select **Send MDN** if required.

11. Click on **Send Settings**:

    °  The **Inbound URL** is where Fabrikam would send messages to reach Contoso. The <Agreement ID> in the URL would be filled in after the agreement is deployed. The inbound refers to the endpoint where messages need to be delivered from the Fabrikam system in order to be sent to Contoso over AS2.

    °  Under **Protocol**, similar to **Send Settings**, configure signing and encryption options if required. Also, set **Acknowledgement** and select **Request MDN** (you can leave the other two checkboxes unchecked).

    °  Under **Transport**, enter the URL of the Contoso endpoint. You can create a BizTalk Server or BizTalk Services setup which can mock the other end of the partner, or simply leave the URL to the default http://www.microsoft.com. Note that the prefix http is required.

12. Click on **Deploy**. You should see the following message in the portal:

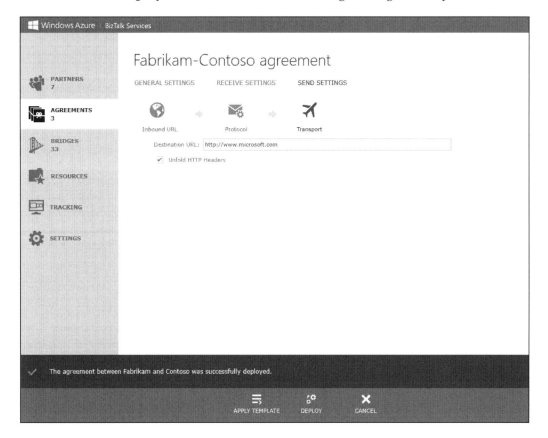

# Configuring the X12 agreement between Northwind and Contoso

Similar to the AS2 agreement, we now need to configure the X12 agreement between Northwind and Contoso. The following steps need to be carried out:

1. Click on **AGREEMENTS** in the left navigation bar.

2. Click on **EDI** this time; the (0) indicates there are no X12 or EDIFACT agreements.

3. Click on **Add**.

4. In the **New Agreement** option, choose the protocol as **X12** and fill the agreement name as `Northwind-Contoso agreement`, and also add a description. You can also create an EDIFACT agreement. The next set of steps configure the specific X12 settings.

5. Select **Hosted Partner** as `Northwind` and **Guest Partner** as `Contoso`. Here, **Hosted Partner** is the partner who directly/indirectly owns the bridge in BizTalk Services.

6. Select **Qualifier** as **ZZ - Mutually Defined (X12)** and enter the value for Northwind as `northwind` and for Contoso as `contoso`.

7. Enable tracking and archiving on both sides; the latter is available for Premium SKUs.

8. Click on **Continue**.

9. The page is now renamed to `Fabrikam-Contoso agreement` and you are on **Receive Settings**.

10. The following are the steps to configure **Receive Settings**:

    ○ Under **Transport**, choose **Transport type** as **AS2** and **AS2 Agreement** as `Fabrikam-Contoso agreement`. This means all messages from the X12 agreement will be received from the AS2 channel configured earlier. In future, `Fabrikam` can create multiple such X12 agreements and reuse the same AS2 agreement to connect with the retailer `Contoso`. This is possible as only the X12 configuration changes across suppliers while the `Fabrikam` connectivity to `Contoso` is mostly constant.

    ○ Under **Protocol**, select **TA1 expected** and **997 expected**. Assuming you have downloaded the B2B schemas from the BizTalk Services download page, click on **Upload** and add `X12_00401_810.xsd` and `X12_00401_850.xsd` one after the other. They will be listed post upload, as shown in the following screenshot:

11. Leave the **Transform** settings as they are for now.

12. Under **Route**, click on **Add** to add the success route rule:

    ° Enter the rule name as `default`.

    ° Click on **Advanced Settings** and enter `1=1` in the expression window. This means we will route all successful messages to this endpoint.

    ° Select **Transport type** as **Azure Service Bus**.

    ° Select **Route destination type** as **BasicHttpRelay**. Add the **BasicHttpRelay URL**, **issuer name**, and **issuer key**. This is the URL where the BasicHttpRelay service will be listening for successful messages routed from this agreement.

    ° Click on **Save**.

13. Add the **Message Suspension Settings**. The suspend settings refer to the target URL if messages fail to successfully reach Northwind. We can configure Azure Service Bus with Queue, Topic, or Relay. Queues and Topics need to be pre created with Shared Access Signature or Issuer Name and Secret:

    ° Select **Transport type** as **Azure Service Bus**.

    ° Select **Route destination type** as **BasicHttpRelay**. Add the **BasicHttpRelay URL**, **issuer name**, and **issuer key**. This is the URL where the BasicHttpRelay service will be listening for failed messages from this agreement.

    ° Click on **Save**.

14. Click on **Send Settings**:

    ° Under **Inbound URL**, note the **Endpoint** value—the `<Agreement-ID>` in the URL will be updated after the agreement is deployed. We will return to this view later.

    ° Leave the **Transform** settings on the send side as the are for now.

    ° Under **Batching**, you can configure to send a batch of messages instead of a single message.

    ° Click on **Add Batch**, enter `batch of 3 messages` as the name, and add a description. Click on **Next**.

    ° In **Batch criteria**, choose **Use advanced definitions** and enter `1=1` in the textbox. This implies all messages will be batches. Click on **Next**.

    ° In **Batch release criteria**, choose **MessageCountBased** and enter `3` as the value of **Count**.

    ° Click on **Next,** and finally, **Save**.

    ° You don't have to click on **Start Batch**; the batch will be started automatically once the agreement is deployed.

15. Under **Protocol**, select **TA1 expected** and **997 expected**. Also, click on the **+** symbol under the **Schemas** section and choose the existing **X12_00401_810.xsd** option from the **Schema** dropdown. This should already be listed if the same was uploaded in the previous **Receive Settings** section.

16. Under **Transport**, fill both the **Transport Settings** and **Message Suspension Settings**. Under **Transport**, choose **Transport type** as **AS2** and **AS2 Agreement** as `Fabrikam-Contoso agreement`. This means all messages from the X12 agreement will be sent to the AS2 channel configured earlier.

17. Add the **Message Suspension Settings**. The suspend settings refer to the target URL if messages fail to reach Contoso. We can configure Azure Service Bus with Queue, Topic, or Relay. Queues and Topics need to be precreated with Shared Access Signature or Issuer Name and Secret. In this case, we configure a service running with BasicHttpRelay binding:

    ○ Select **Transport type** as **Azure Service Bus**.

    ○ Select **Route destination type** as **BasicHttpRelay**.

    ○ Add the **BasicHttpRelay URL**, **issuer name**, and **issuer key**. This is the URL where the BasicHttpRelay service will be listening for messages which could not be sent to `Contoso`.

18. Click on **Deploy**:

    ○ A success message, `The agreement between Northwind and Contoso was successfully deployed` should be shown. If not, check for errors and click on **Deploy** again.

    ○ Navigate to the agreement's **Send Settings** page and note down the **Inbound URL** value. This will be used to send messages to the agreement.

# Sending messages

**Web Sender** is a tool available in MSDN Code Gallery for BizTalk Services' samples. It can be used to send messages to AS2 receive endpoints. Since in this example the AS2 is tied to X12 endpoints, all messages sent to AS2 receive should be routed to the success endpoint of X12 receive (the Azure Service Bus relay endpoint in this case). **Message Receiver** is a tool in MSDN Code Gallery to receive messages on relay endpoints; it needs to listen on the configured address in the X12 agreement route address.

Use **Web Sender** from MSDN Code Gallery to test the AS2 message with sample 850. You can generate an instance of the Purchase Order EDI 850 using Visual Studio **Generate Instance** command from the BizTalk Services project.

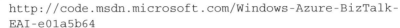

You can download the Message Receiver C# sample at the following link:

`http://code.msdn.microsoft.com/Windows-Azure-BizTalk-EAI-e01a5b64`

You can download the Web Sender C# sample at the following link:

`http://code.msdn.microsoft.com/Windows-Azure-BizTalk-a0d12dca`

## Viewing tracking data

Click on the **TRACKING** view in the navigation bar to see the status of message flow on the agreement.

# Summary

In this chapter, we started with the basic concepts around B2B and its relevance in the context of Azure. We introduced key concepts of B2B on Azure, notably partners, agreements, templates, batching, tracking, and archiving. We also walked through a simple agreement configuration for AS2 and X12 in BizTalk Services. Trading partners can also be managed using API—this is covered in detail in the extensibility chapter later.

# 6
## API

So far, we've only looked at the graphical tools to interact with **Windows Azure BizTalk Services (WABS)**. These have included Visual Studio for creating and deploying solutions as well as the BizTalk Services Portal (and Azure Management Portal) for management and monitoring of the deployed solutions. Underlying all these tools though is a REST-based API that allows easy integration with scripting tools as well as your own processes to facilitate automated actions such as deployment, testing, and management.

In this chapter, we'll look at the WABS API and how to interact with it using:

- RESTful web services
- PowerShell
- Custom code

While there are three methods in which the API can be used (portal, REST services, and PowerShell), each serves a different need, and although there is an overlap, there are also differences in functionality. The portal has been explored elsewhere in this book and provides an easy-to-access dashboard for system administrators. PowerShell is a familiar tool for IT pros that is useful for scripting system interactions such as deployments. Using the REST API directly is useful for building your own tools and capabilities on top of WABS or for interacting with WABS from another application. By the end of this chapter, you'll have a good understanding of the WABS API and how you can leverage it to your advantage in your own organization.

# REST

First, let's take a quick look at the grounding of the provided API. All functions available in Visual Studio and in the management portal are also available in the API. In fact, the API actually provides more capabilities than these tools do, as we'll see. This shouldn't be too much of a surprise as it is often the case—the API usually comes first and the tools later. It's therefore a good idea to understand what the API can do. Underpinning this API is a set of web services accessible using HTTP. WABS uses RESTful services for this. REST is not a standard or protocol, but an architectural style that enables simple HTTP-based integration. It doesn't need the overhead of SOAP or frameworks such as Microsoft's WCF. In fact, you can often use just your web browser to make requests or query for information. REST is based on a set of standard HTTP verbs that specify the type of request. WABS uses the following HTTP verbs in its API:

| Verb | Purpose |
| --- | --- |
| PUT | Create a new artifact or update an existing one |
| GET | Retrieve artifact(s) |
| DELETE | Remove an artifact from WABS |
| POST | Update an artifact or service status |

WABS REST API verbs

As you can see, the full set of CRUD (create, read, update, and delete) operations are supported in this way, which provides a great deal of flexibility as it facilitates cross-platform access and easy integration with third-party tools.

# Calling the API

Let's start by looking at a simple REST call to the BizTalk Services API. In this example, we'll query the BizTalk Services instances deployed for a given Azure subscription. We are going to see how you can execute this request using a very useful tool called Fiddler. You can download Fiddler for free from http://fiddler2.com/.

In order to execute these API calls against Azure, a mutual certificate exchange process is required in order for each party to authenticate one another. When your machine makes a request to the Windows Azure management endpoint, Azure returns the configured certificate and in return your client machine sends Azure its certificate to validate. Once complete, Azure executes the request and returns an acknowledgement. In order for this to work, we first need to create a client certificate and then upload it to Azure.

There are two options here. You can use a certificate you create yourself, which is known as a self-signed certificate. Such a certificate is useful for testing but would not be appropriate for the production usage. In this case, you would purchase a certificate from a signing authority and use that. The reason for this is that certificates are about trust, not just between the two parties (your machine/organization and Azure), but with the signing authority as well. When a party receives a certificate, it can check its validity with the signing authority. This also allows, for example, the ability for a signing authority to revoke a certificate if it has been compromised.

For our purposes though, a self-signed certificate will do just fine. To create a certificate, open a command prompt and enter the following command:

```
makecert -sky exchange -r -n "CN=wabstest" -pe -a sha1 -len 2048 -ss
My "%HOMEPATH%\documents\wabstest.cer"
```

This command will create a self-signed certificate and install it in your machine's certificate store under your logged on account. With this done, we need to associate it with our Azure subscription where we have provisioned BizTalk Services:

1. Open the Windows Azure Management Portal at `http://manage.windowsazure.com`.

2. In the left-hand margin, click on **Settings** (it's the last one in the list).

3. Under **Settings**, click on the **Management Certificates** tab and then click on **Upload**.

4. Browse to the certificate file you created in the command window earlier—it will be located by default under `c:\users\<youraccount>\documents`.

5. Click on the tick button to associate your certificate with the management service.

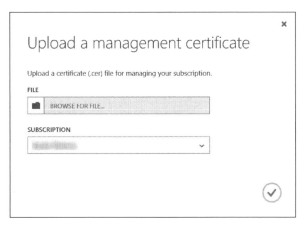

Upload management certificate

Now that we've done this, we can make a call using the Request Composer feature of Fiddler to query the WABS services deployment. To set up the certificate in Fiddler, we need to perform a few steps first before making the request:

1. Open Fiddler.

2. On the **Rules** menu, select **Customize Rules...**.

3. In the `CustomRules.js` file, which opens in Notepad, find the `OnBeforeRequest` function.

4. Add the following at the top of this function, replacing `<username>` with your username:

```
if (oSession.HostnameIs("management.core.windows.net")) {
    oSession["https-Client-Certificate"] = "C:\\Users\\<username>\\
Documents\\wabstest.cer";
}
```

5. Save the file and close Notepad.

What this will do is send the client certificate to the service whenever the Azure management URL is accessed. For the next step, you will need your Azure subscription ID. To get this, go back to the Azure Management Portal, and under **Settings | Subscriptions**, you will see a list of your subscriptions in the **Subscription** column and the required subscription IDs in the **SubscriptionID** column, as shown in the following screenshot:

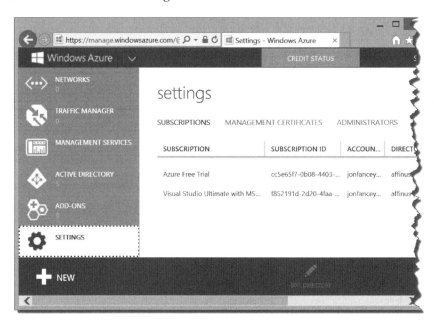

Obtaining the subscription ID

Now, we can make the request as follows:

1. Click on the **Composer** tab.

2. Ensure the verb next to the URL is set to **GET**.

3. Enter the following URL in the box, replacing `<SubscriptionID>` with your own:

   ```
   https://management.core.windows.net/<SubscriptionID>/
   cloudservices
   ```

4. Add the header provided here in the **Request Headers** area:

   ```
   x-ms-version:2010-10-28
   ```

   This HTTP header specifies the service version we want and is mandatory. Currently, there is only one version, but over time the service may change, and this will allow you to call a particular version of it.

   Your Fiddler request should look like the following screenshot:

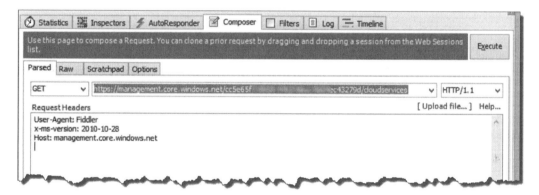

Retrieving a list of cloud services

5. Click on the **Execute** button.

If all went as planned, you should now see the results of the call in the Fiddler window as shown in the following screenshot. What you are looking at is a list, in XML format, of all the BizTalk Services instances provisioned for the subscription you passed in as the argument. If you were to call this API programmatically, you could read through the XML and pull out particular properties for each instance and perhaps stop or restart them all. I've blurred out the subscription IDs and other details for obvious reasons.

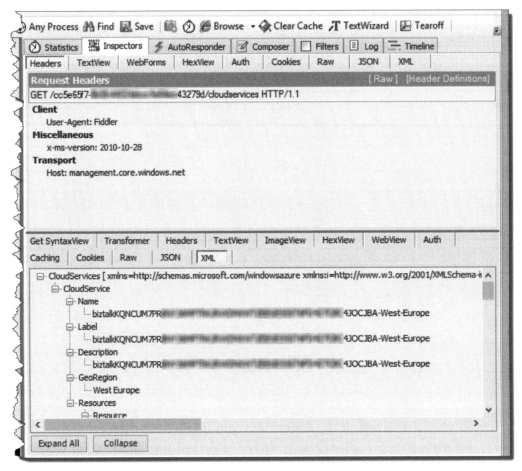

Retrieving a list of cloud services

With the results of this call, we can now retrieve the details of a single WABS instance with the following URL. Here, the cloud service name returned by the previous call is passed into the Get Cloud Service call:

```
https://management.core.windows.net/<SubscriptionID>/cloudservices/<C
loudServiceName>
```

The request and response are shown in the following screenshot:

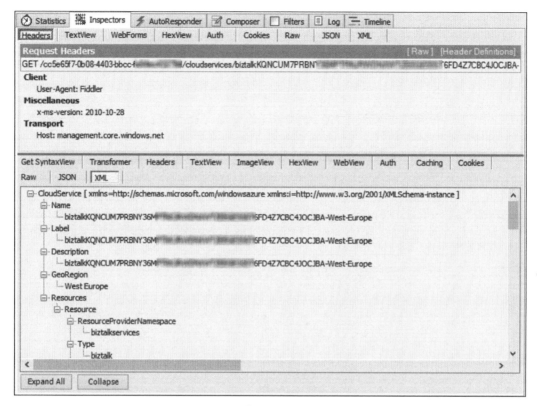

Retrieving a single BizTalk Services instance

# Back up and restore

Now that we've looked at a simple example of what the WABS API can do, let's look at some of the more interesting capabilities. An essential aspect of enterprise development is the ability to move artifacts between environments. Commonly, an organization or team will have a development, testing, user acceptance, and production environment (and multiple instances of each). This DTAP (dev, test, user, and prod) setup is perfectly possible with BizTalk Services by creating multiple service instances and provisioning them as required. Then, each can be used as desired to manage the overall integration estate.

Backing up a BizTalk Services instance is not just useful to move content between environments, but can also be used to keep a set of backups or snapshots of a particular environment for disaster recovery or to restore to a particular point in time. It is also possible to restore an instance to a different version of the service, provided that the service type is at least the same or higher. For example, a Basic subscription can be restored to not just another Basic sub, but to Standard or Premium as well. Downgrading however, is not possible, and nor is backing up a Developer instance of the service.

This feature now has (as of the February 2014 service update) out of the box tooling via the Windows Azure Management Portal as shown in the following screenshot in the **CONFIGURE** tab. While the portal UI now allows you to back up a service instance and even create a new BizTalk Services instance from a backup, using the API programmatically is very useful. The API provides the ability to move or "promote" a set of artifacts from one instance (say Test) to another (for example, User Acceptance) programmatically. In this section, we'll look at how to achieve this capability with the REST API by writing some .NET code to do it. As you'll see, this is very easy and straightforward to perform.

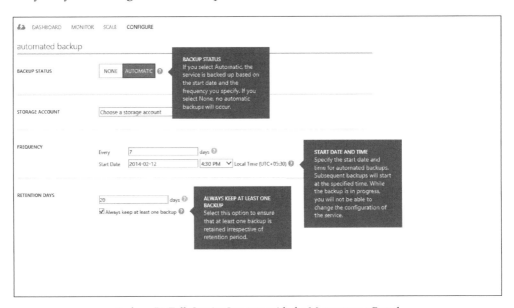

Back up BizTalk Service Instance with the Management Portal

Before trying this out, I should point out that this capability provides a similar copy of the service instance. It's quite possible (even likely) that some of your settings or configuration is environment specific. For example, if your bridges send messages to Service Bus queues, it is unlikely that you would use the same queues for test as production. Therefore, while being able to back up one environment and restore to another is certainly very useful, you also need to think about using the REST API to apply configuration changes on top of a restored service instance.

Open Visual Studio and create a new console application. Call it `BackupService`. In the static `Main` method, add the following code to replace the empty `Main` method:

```
static void Main()
  {
     Task t = new Task(Run);
     t.Start();
     Console.ReadLine();
  }
```

Now add the `Run` method as shown in the next code snippet. This code formats the required URL to make the backup API call. For this, three pieces of information are required.

Firstly, you'll need your Windows Azure subscription ID; this is the same ID as discussed earlier, and you can obtain it as before through the Windows Azure portal in the *Upload management certificate* screenshot in the *Calling the API* section. You'll also need the service name. This is the value in the **Name** field as shown previously, and you can obtain yours by making that API call in Fiddler as we saw. The final piece of data you need is the resource name of the BizTalk Services instance. This is the name you gave your WABS instance when you created it. It can be obtained either through the Azure portal, by clicking the BizTalk Services link, or again by using Fiddler as shown in the previous screenshot. The name you need is under the **Resources/Resource/Name** element. Replace the three placeholders in the code with your service values as shown in the following code:

```
static void Run()
{
   string subscriptionId = "<SubscriptionID>";
   string operationName = "cloudservices";
   string serviceName = "<servicename>";
   string resourceName = "<resourcename>";
   Uri requestUri = new Uri("https://management.core.windows.net/"
                               + subscriptionId
                               + "/cloudservices/"
                               + serviceName
```

```
                                          + "/resources/biztalkservices/~/
biztalk/"

                                          + resourceName
                                          + "/?comp=backup");
    MakeRequest(requestUri);
}
```

Now, add the following two assembly references that contain the necessary types to make the request to the service endpoint:

```
System.Net.Http
System.net.Http.Request
```

Add the following method directly underneath the previously added code. This will set up the call to the Backup REST API, and to do this, it needs your certificate. As discussed earlier, the management API calls authenticate services using mutual certificates, and therefore we need to pass our certificate. However, as the rule that we added to Fiddler earlier sends the client certificate with each request made to the management URL, we don't need to send the certificate in the code—you just need to ensure that Fiddler is still running (the code is provided to add the certificate later if you want to run it without Fiddler). This simplifies things quite a bit.

```
static async void MakeRequest(Uri requestUri) {
    string payload =
"{\"BackupName\":\"<backupname>\",\"BackupStoreConnectionString\":\"Ac
countName=<storageaccountname>;AccountKey=<storageaccountkey>;DefaultE
ndpointsProtocol=https\"}";

    HttpContent content = new StringContent(payload);
    content.Headers.ContentType.MediaType = "application/json";
    content.Headers.Add("x-ms-version", "2010-10-28");
    using (var client = new HttpClient())
    {
        var response = await client.PostAsync(requestUri, content);
        response.EnsureSuccessStatusCode();
        Console.WriteLine("Backup started");
    }
}
```

You need to replace the `<storageaccountname>` and `<storageaccountkey>` values in the preceding code with your own storage account details. To obtain your `AccountName` and `AccountKey` values, do the following:

1. Go to the Azure Management Portal.

2. Click on the Storage icon in the left-hand navigation bar.

3. In the list of storage accounts, select the one that has the same name as the one created by your BizTalk Services instance.

4. Click on the **Manage Access Keys** button at the bottom of the page.

5. Copy and paste the **Storage Account Name** and **Primary Access Key** fields into the preceding code.

You can actually use any storage account you like in step 3, or even create a new one. The account is used to store the backed-up WABS instance. The third placeholder in the code is `<backupname>`. This is the label to use for backup, and it is good practice to name this something meaningful, such as with the date the backup was made. The label you use must start with a letter or a digit, can only contain digits, dashes (-), or lowercase letters, and can be between 3 and 63 characters long. Dashes must not be consecutive.

The format of the data posted to the service is **JSON (JavaScript Object Notation)**, and this is simply a string containing the storage account details and backup name. The PostAsync call will invoke the API and wait for a response. If successful, the service will return an OK response HTTP code 200. This API is asynchronous because the service backup can take up to an hour to complete. In response, we get a tracking identifier that allows you to check the status of the backup operation. The API provides a polling query for this that allows you to make a call passing your returned identifier (a GUID) and retrieve the results of the operation at any point. In this way, you can ensure that the backup was successfully completed.

Now that the code is complete, hit *F5* to build and run it. If successful, the console application should just open and close down a few seconds later. You might want to put a couple of breakpoints in the code and run it to see if it's working. I've also omitted any exception handling code for brevity. If it fails to call the API, an exception will be thrown. In such a case, run the debugger so you can determine what the problem is.

Of course, there is also a reciprocal Restore API call that allows you to restore a previously backed up instance to any other BizTalk Service instance.

As I mentioned previously, you need to keep Fiddler running for this example as Fiddler is supplying the necessary certificate. If you want to run without Fiddler, just add the following code at the start of the `MakeRequest` method, replacing the `<your thumbprint>` placeholder with your own certificate's thumbprint as displayed in the Azure Management Portal:

```
var certHandler = new WebRequestHandler();
string certThumbprint = "<your thumbprint>";
X509Store certStore = new X509Store(StoreName.My,
                                    StoreLocation.CurrentUser);
certStore.Open(OpenFlags.ReadOnly);
X509Certificate2Collection certCollection = certStore.Certificates.
Find(X509FindType.FindByThumbprint, certThumbprint, false);
certStore.Close();
X509Certificate2 certificate = certCollection[0];
certHandler.ClientCertificates.Add(certificate);You also need to
change the using statement as shown below to pass in the certificate
from:
using (var client = new HttpClient())
To:
using (var client = new HttpClient(certHandler))
```

The preceding code retrieves your certificate from your local machine's certificate store. You therefore need to ensure it is stored already. To do this, double-click on your certificate, and in the wizard that appears, do the following:

1. Accept any security warnings first.
2. Click on the **Install Certificate** button.
3. For the **Store Location** option, select **Local Machine**.
4. Accept any warning that appears.
5. Select **Place all certificates in the following store**.
6. Click on the **Browse** button.
7. Select **Personal** and click on **OK**.
8. Click on **Next** and then **Finish**.
9. You should see a message confirming successful installation.
10. Close the dialog.

# Using PowerShell

So far, we've seen two different ways to utilize the API provided by BizTalk Services, directly making HTTP requests in Fiddler and by writing code to make the calls to it programmatically. Now we'll look at an even easier way, using Windows PowerShell. Windows PowerShell is a command-line tool aimed at administrators that provides a consistent way to perform tasks across many Microsoft products (and third-party ones). With PowerShell, it is possible to automate common actions and create sophisticated scripts that perform configuration and administration of BizTalk Services environments and Azure in general.

BizTalk Services provides a set of PowerShell cmdlets that can call the complete set of APIs provided. Cmdlets are units of functionality that are executed in PowerShell, and BizTalk Services provides a cmdlet for each API call available.

To be absolutely correct, BizTalk Services actually provides two sets of cmdlets. The first is installed when you download and install the BizTalk Services SDK while the second needs to be downloaded. The first allows control over the artifacts in a provisioned BizTalk Services instance while the second allows control over BizTalk Services as a whole—including creating new BizTalk Services instances. As the second set is associated with the APIs we've already been looking at, we'll start there. This second set is provided as source code and can be downloaded from the following link:

```
http://code.msdn.microsoft.com/windowsazure/Windows-Azure-BizTalk-
91e1bdf3
```

As it is the source code, it needs to be opened and compiled in Visual Studio. We should also note that this is a sample and not officially supported code from Microsoft. Once the source code is built, open PowerShell on Windows 8/Server 2012 by clicking on the Start button and typing `PowerShell` (on Windows 8 or 2012). You should see Windows Azure PowerShell appear in the list of results. Click on it to launch it. If you don't see Windows Azure PowerShell, make sure you have installed it and you have at least Version 0.6.19 installed.

In the PowerShell command window, enter the following command to load the cmdlets:

```
import-module
  <pathtosource>/Microsoft.WindowsAzure.Management.BizTalkService.dll
```

In order to use the cmdlets, the subscription context must first be set. Do this by entering the following code in the command window:

```
$sub = '<subscription ID>'

$thumbprint = '<certificate thumbprint>'

$cert = Get-Item cert:\\LocalMachine\My\$thumbprint

Set-AzureSubscription -SubscriptionName "Test" -SubscriptionId $sub
-Certificate $cert

select-azuresubscription –SubscriptionName "Test"
```

You should now know how to obtain the value for `<subscription ID>` that needs to be substituted. For `<certificate thumbprint>`, if you followed the steps earlier to generate and upload a certificate, you need to replace this value with the thumbprint of your own certificate. To find this, go to the Azure portal and click on **Settings** in the left-hand navigation bar. On the Settings page, click on **Management Certificates** and then cut and paste the value for the thumbprint column for the certificate you uploaded earlier.

I've used the value of `Test` in the previous code to name the subscription. This can be any label you like. It is only used to name the subscription during the PowerShell session. Now, as soon as this is done, all cmdlets will be executed in the context of the particular subscription.

As an example of how to use the cmdlets, let's look at one of the API calls we made earlier. In the command window, type the following command, substituting the name of your BizTalk Service instance for `<service name>`:

```
Get-AzureBizTalkService -resourcename <service name>
```

You should see a response in the command window similar to the one in the following screenshot:

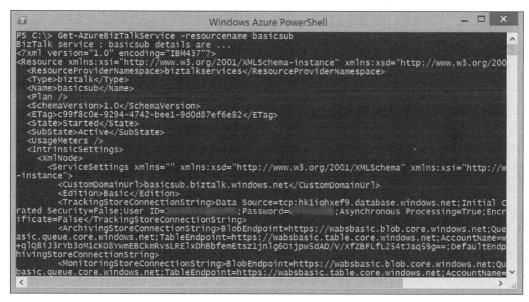

Get BizTalk Service cmdlet

Using the API doesn't stop at just being able to query the service. We can also create a brand new BizTalk Services instance or delete an existing one. It is also possible to suspend or resume a particular service instance if required. To create a new instance, the `New-AzureBizTalkService` cmdlet is provided. This takes the following form:

```
New-AzureBizTalkService -ResourceName MyNewBizTalk -Location "West
Europe" -ConfigurationFile "c: \ create_new.xml"
```

Apart from the name of the instance and what data center to create it in, the main parameter is actually a file. The download for the source code actually contains a couple of example files that you can adapt for this purpose. The file you provide contains all the details that you would normally specify when creating a new service via the Azure portal; for example, the database to use, the certificate to protect the service with, the type of service—developer, premium, among others—and the ACS settings. Given what you know after reading this book, you should find editing the provided sample files with your settings quite straightforward. Once done, you can automate the creation of services to your heart's content!

OK, so far we've covered the management aspects of the BizTalk Services API. But as mentioned earlier, there is also another set of PowerShell cmdlets that are used to manipulate artifacts and settings in a BizTalk Services instance. This set of cmdlets is already installed if you have the BizTalk Services SDK installed, which by default is located under `C:\Program Files\Windows Azure BizTalk Services Tools`.

To load the cmdlets, type the following command in the PowerShell window:

```
import-module "C:\Program Files\Windows Azure BizTalk Services Tools\
Microsoft.BizTalk.Services.Powershell.dll"
```

This PowerShell module provides features that are not available in the UI of BizTalk Services via the portal. One example is the ability to start and stop a bridge. When a new bridge is deployed, it is active by default, but there are times when you may wish to stop a bridge from receiving messages. This can be achieved with the `Stop-AzureBizTalkBridgeSource` cmdlet as follows:

```
Stop-AzureBizTalkBridgeSource –AcsNamespace <namespace> –IssuerName owner
–IssuerKey <key> –BridgePath MyBridge
```

This will stop all sources available on the bridge `MyBridge`, but it's also possible to stop a particular source by providing the `SourceName` parameter. This is very useful when you need to perform maintenance that requires some or all of the sources to be temporarily stopped. To restart a bridge/source, the corresponding `Start-AzureBizTalkBridgeSource` cmdlet is used with the same parameters.

The remaining cmdlets concern adding and removing artifacts such as bridges, schemas, certificates, and assemblies to a BizTalk Service. Visual Studio uses these API calls during deployment and their primary usage outside of this is to automate and manage deployments. For the full list of cmdlets, visit `http://msdn.microsoft.com/en-us/library/windowsazure/dn232360.aspx`.

# Summary

In this chapter, we have looked at the API underpinning BizTalk Services. We've seen how to leverage the API from the humble web browser and how to use PowerShell cmdlets and write our own code to invoke it. We've looked at the different types of APIs, capabilities, and the cmdlets that wrap all of them, and hopefully you've seen how you can make use of the capabilities of the BizTalk Services API to create, manage, maintain, and more importantly, automate your BizTalk Services instances. In the next chapter, we will look at troubleshooting your integration solutions and how to use the tracking capabilities of WABS.

# 7
# Tracking and Troubleshooting

In the last few chapters, we looked at the artifacts used in building a BizTalk Services solution. By now, you must be wondering how to track the message flow or, still better, how to troubleshoot if things didn't go as expected. In this chapter, we will look at the tools and common patterns to troubleshoot issues in BizTalk Services.

Specifically, we will focus on troubleshooting the following topics:

- Sources and destinations
- Schemas and transforms
- EAI bridges with custom code
- B2B agreements
- Hybrid connectivity

## Messages and errors

First, let's quickly summarize the basics. Bridges are message channels that don't persist messages. This means any failure in the message processing will be returned as an HTTP error to the caller in the case of EAI bridges, and the message will be pushed to the suspend endpoint in the case of B2B bridges. The suspend endpoint is important in the case of B2B as the error in configuration or message structure cannot be sent back to the business partner, but is meant for consumption by the IT operator. This means in both EAI and B2B scenarios, all retries and resubmissions of messages post failures have to be done outside the bridge.

Errors can occur in any of the following three scenarios:

- **Errors during deployment time**: This scenario includes all the errors associated with provisioning of the BizTalk Service deployment. In most cases, the error is self-explanatory and is shown in the Windows Azure Management Portal or sent back via the RDFE API. It is important to note that a BizTalk Service deployment name is unique. Custom domain merely serves to wrap a DNS name around the BizTalk Service deployment URL. The certificate of the domain needs to be uploaded in the Trusted Root Certification Authorities certificate store on the machine accessing the deployment. The storage and the Azure SQL Database used for tracking and archiving cannot be reused or deleted while the deployment is active.

- **Errors during design time**: This scenario includes all the errors during adding/updating/deleting a bridge, deploying a VS project, or adding/updating/deleting an agreement from the BizTalk Services Portal. These errors surface in the **Output** window or the **Error List** window in Visual Studio for the EAI scenarios and in the status bar of the BizTalk Services Portal for the B2B scenarios, respectively.

- **Errors during runtime**: This scenario includes all the errors during the actual flow of messages between two applications or partners. This scenario can be further broken down into four subcategories:
  - Errors when the message is sent to an endpoint external to BizTalk Services is faulty.
  - Errors when the message is expected to be received from an endpoint external to BizTalk Services is faulty.
  - Errors when the message is malformed and does not conform to the schemas configured in the bridge.
  - Errors when components such as the bridge, transform, source, or hybrid destination in BizTalk Services do not function as expected. This usually classifies a bug in the product.

We will focus on runtime errors and the first three subcategories in particular. If tracking is enabled, tracking records are logged in the Windows Azure SQL Database. Tracking enables us to store interesting properties related to the message—from the header, body, or through lookup from another data source. Archiving persists the message data in a raw form in the case of EDI scenarios. For the EAI scenarios, archiving is possible by adding custom code as outlined in *Chapter 4, Enterprise Application Integration*. Data written into the Tracking and Archiving stores is carried out on a *best effort* basis, that is, if there is an error during the write operation to these stores, tracking and archiving will be skipped and message processing will continue in the bridge. The exception to this case is when the archiving of the AS2 messages is enabled with the **Enable NRR** option turned on in the AS2 agreement **General Settings** page. In these cases, the message processing fails if tracking/archiving cannot be completed successfully.

# Data for troubleshooting

In this section, we'll explore the different kinds of data available to troubleshoot issues.

# Tracking

Every message that flows through the bridge is associated with a promoted property known as the **Request ID**, which is a GUID value on each incoming message. If the message is split into submessages, each submessage gets its own tracking ID, which is also a GUID. If the Request ID is the same as the tracking ID, the message flows without debatching. The bridge endpoint URI and timestamp should point to the bridge and timing of the message. Tracking can be enabled from the bridge properties in VS and from an agreement's **General Settings** page in the BizTalk Services Portal.

The BizTalk Services Portal exposes the tracking data in a user-friendly way. There are three tabs that reflect the messages processed in the deployment. They are explained as follows:

- **MESSAGES**: This tab contains all the messages from sources, bridges, and agreements with errors or information-type entries. Each tracking entry details the message's incoming URL, its Request ID and tracking ID, whether the processing was an error or a success, the stage where the track record was emitted, and the date and time when it occurred. Use this view for tracking all the EAI and B2B messages passing through bridges and agreements.

- **PROTOCOL**: This tab lists the track records for B2B interactions. The view is also categorized into the EDI and AS2 protocol levels. EDI calls out a message status for X12 and EDIFACT records with sender, receiver, message type (such as PO), acknowledgments such as technical acknowledgment and functional acknowledgment, Request ID, ID of the interchanged envelope to correlate with the track records of batching, and the date and time when this record was written. The AS2 records contain similar information, except that the acknowledgment reflects the **Message Disposition Notification** (**MDN**) status. Use this view to track all the B2B protocol stage-specific tracking entries.

- **BATCHING**: Finally, the **BATCHING** tab tracks the list of ongoing and completed batches along with the individual message information. The view tracks the batch name, the agreement for which the batch is configured, and the sender and receiver of the batching transaction. The entry also shows the size, count, and time when the transaction was received using which the customer can relate to the expected release criteria of the batch. Use this view to track all the messages in a batch.

Each of the tabs also has a **Search** option, which can help filter the result by date range, message type, status, sender, or receiver.

 Without searching for any option, the search option displays all the track records sorted by the latest date. This can also be used to refresh the page during testing.

The **Tracking** view is shown in the following screenshot:

Tracking view in the BizTalk Services Portal

In some cases, it is required that you access the data directly from the Azure SQL Database tables. A common use case might be to build a notification system based on tracking events. The Azure SQL Database tracking tables used in tracking are as follows (note that none of these tables are supported or documented from Microsoft for issuing direct T-SQL queries):

| Table | Data |
|---|---|
| [dbo].[PipelineTrackRecords] | Bridges tracking records |
| [dbo].[SourceTrackRecords] | Sources tracking records |
| [dbo].[EndpointAddressMap] | Stores the URL of a bridge or a source and maps the address to the Pipeline and Source track records using a foreign key reference, EndpointAddressID |
| [dbo].[TrackRecordMessageProperties] | Name and value pair of promoted properties of the message |
| [dbo].AS2*, [dbo].Batch*, [dbo].Functional*, [dbo].Interchange*, [dbo].TransactionSet* records | EDI tracking records for AS2, MDN, batching, Interchange, Group, Transaction set records, and functional or technical acknowledgments |

# Traces and logfiles

Besides tracking, trace statements are also recorded in the Azure tables as a message flows through a bridge. Traces are useful to look for exceptions when the time period of the message failure is known. The information from tracing can supplement the tracking information from the BizTalk Services Portal. These traces are similar to the **Event Trace Log** (ETL) traces, except that the BizTalk Services traces are text based and stored in the Azure tables.

For each deployment, traces are logged on the Azure table named **WADLogsTable** created in the storage account specified while provisioning the BizTalk Services deployment. You can use a tool such as **Azure Storage Explorer** from `azurestorageexplorer.codeplex.com` or one of the commercial tools such as CloudBerry to connect to that Azure storage account and view the data in the table.

The following three fields in WADLogsTable are interesting:

- **Timestamp**: The date and time when the traces were logged.
- **Message**: Information, exception, or error message with a component or activity information.
- **Level**: Trace level varying among Fatal (1), Errors (2), Warnings (3), and Informational (4). The errors in level 2 are accompanied by the exception stack trace.

Traces are extremely useful when troubleshooting custom code configured with bridges. As there can be hundreds of entries in a few minutes, you can filter the data in the table using one of the following commands in the **Azure Storage Explorer** tool:

- `Timestamp gt datetime'2013-12-07T16:00:00'`
- `Level = 2`

Note that the spacing as well as the casing is important in filtering the data.

In the case of the BizTalk Adapter Service, logfiles can be written by adding log interceptors in the service `.config` file. To troubleshoot the hybrid connectivity runtime, edit `web.config` in `C:\Program Files\Microsoft BizTalk Adapter Service\BAServiceRuntime`. The exact entries that must be added to generate logfiles are outlined in the *Troubleshooting hybrid connectivity* section.

# Performance counters

You can use performance counters to assess the health of the system. Performance counters pertaining to the BizTalk Services deployment are stored in the storage account of the deployment and can be viewed from the Azure Management Portal's **MONITOR** tab as shown in the following screenshot:

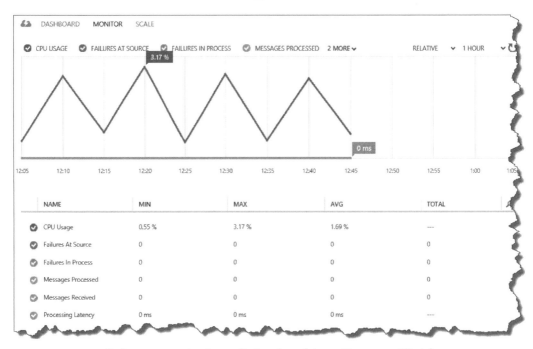

Performance counters in monitoring view of Azure Management Portal

The following performance counters are available for each deployment:

| Performance counter name | Unit | Description |
| --- | --- | --- |
| CPU Usage | % | Average CPU usage of all instances servicing the runtime messages |
| Failures at Source | count | Count of messages that failed in the sources |
| Failures in Process | count | Count of messages that failed during pipeline processing |
| Messages in Process | count | Count of messages currently in process by the deployment |
| Messages Processed | count | Count of messages successfully processed by the deployment |
| Messages Received | count | Count of messages received by the pipelines |
| Messages Sent | count | Count of messages sent or routed from each pipeline |
| Processing Latency | milliseconds | Average time taken to process a message from the validate stage to route for one-way bridges |
| Round Trip Latency | milliseconds | Average time taken to process a message round trip in two-way bridges |

These counters can be useful to make configuration changes to the deployment. For example, if **Messages Received** is trending higher and this correlates with an increase in **Failures in Process** and a corresponding increase in **Processing Latency**, then the system may not be scaling up with the incoming rate. The IT administrator could plan scaling the deployment and look for changes in the performance counters.

# Troubleshooting sources and destinations

Sources can be one of the following: HTTP, FTP(s), SFTP, or Service Bus Queue and Topic. If the source endpoint is HTTP, it is common to see HTTP error codes on the client side sending the message, as shown in the following table:

| Error scenario | HTTP error code | Description |
| --- | --- | --- |
| Message to a nonexistent endpoint or wrong URL | 400, 500 | Bad Request, Internal Server Error, or Namespace cannot be resolved |
| Endpoint with malformed message headers | 401 | Authentication failed or Unauthorized request |
| Endpoint with malformed message body | 500 | Internal Server Error; see tracking or trace entries for more information |
| Destination endpoint down | 500 | Internal Server Error |
| Destination with incorrect credentials | 500 | Internal Server Error |
| Bridge destination is configured for HTTP relay but receiver is listening on HTTPS | 500 | Internal Server Error |

In the case of FTP as source, if there are errors during the processing or at the destination, the message will not be deleted from the source. The polling interval would increase and the system would autoretry the submission of the message.

The following screenshots show the increase in the **NewPollInterval** field as seen in the **PORTAL TRACKING** view for the source name, **Poll Error**. Note that the poll interval increases by 1.5 times the current poll value for each new iteration. The next set of poll intervals would be around 227, 341, 512, and 768 seconds.

| message info | | 23 |
|---|---|---|
| **PROPERTY NAI** | **PROPERTY VALUE** | |
| Error | Could not find file 'C:\Resources\directory\183d1330621449dabbb4160a356695d1.IntegrationRole.TempStorage\Invoice_120.pdf'. | |
| ErrorMessage | Errors were encountered in the FTP poll. Increasing the poll interval to 45 seconds. | |
| EventLevelValue | Error | |
| NewPollInterval | 45000 | |
| State | Faulted | |
| EndpointAddress | /myftpqpassthrough | |
| SourceName | FTPSource1 | |

| message info | | 23 |
|---|---|---|
| **PROPERTY NAI** | **PROPERTY VALUE** | |
| Error | Could not find file 'C:\Resources\directory\183d1330621449dabbb4160a356695d1.IntegrationRole.TempStorage\Invoice_120.pdf'. | |
| ErrorMessage | Errors were encountered in the FTP poll. Increasing the poll interval to 67.5 seconds. | |
| EventLevelValue | Error | |
| NewPollInterval | 67500 | |
| State | Faulted | |
| EndpointAddress | /myftpqpassthrough | |
| SourceName | FTPSource1 | |

| message info | | 23 |
|---|---|---|
| **PROPERTY NAI** | **PROPERTY VALUE** | |
| Error | Could not find file 'C:\Resources\directory\183d1330621449dabbb4160a356695d1.IntegrationRole.TempStorage\Invoice_120.pdf'. | |
| ErrorMessage | Errors were encountered in the FTP poll. Increasing the poll interval to 101.25 seconds. | |
| EventLevelValue | Error | |
| NewPollInterval | 101250 | |
| State | Faulted | |
| EndpointAddress | /myftpqpassthrough | |
| SourceName | FTPSource1 | |

| message info | | 23 |
|---|---|---|
| **PROPERTY NAI** | **PROPERTY VALUE** | |
| ErrorMessage | Errors were encountered in the FTP poll. Increasing the poll interval to 151.875 seconds. | |
| EventLevelValue | Error | |
| Inspector | GeneratePDFInvoice.GenerateSignedPdfInvoice, GeneratePDFInvoice, Version=1.0.0.0, Culture=neutral, PublicKeyToken=773c78de83f452a5 | |
| NewPollInterval | 151875 | |
| State | Faulted | |
| EndpointAddress | /myftpqpassthrough | |
| SourceName | FTPSource1 | |

Tracking entries indicating exponential poll over an FTP source

Some error scenarios while using FTP are as shown in the following table:

| Error scenario | HTTP error code | Description |
|---|---|---|
| Wrong FTP URL or incorrect username or password | 503 | Failed to connect to FTP server and/or not logged in |
| Redeploy FTP while the existing service is active | 400 | One or more resources are in the started state; you can stop the source using `PSCmdlet` |

You can fix the issue based on the error message and use `PSCmdlet` `Stop-AzureBizTalkBridgeSource` and `Start-AzureBizTalkBridgeSource` to stop and start the source respectively. The following screenshot shows the execution of `Get-AzureBizTalkBridgeSource` to check the status of a source endpoint:

Getting source status using PSCmdlet

# Troubleshooting schemas and transforms

Issues in schemas surface when a message fails validation against a schema. If the validation fails, say due to extra tags, then a tracking record is added for the XML validation stage, thus reporting an error.

If the tracking entry indicates a schema validation error, the easiest way to test the schema is to generate a test message. For EAI/B2B schemas, Visual Studio provides a handy utility to generate an instance of the schema. After the schema is added to the project, right-click on the schema and choose **Generate Instance** of the file. From the **Properties** window of the schema, you can generate an instance in native (for flat file) or the XML format, as shown in the following screenshot:

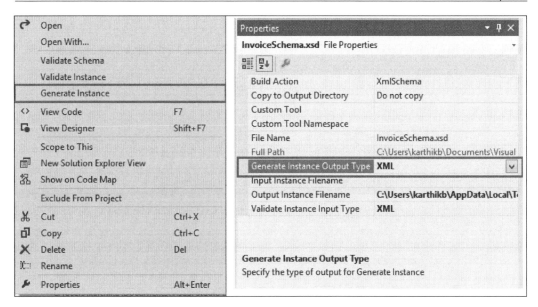

Generating an instance of a schema from Visual Studio

Similar to schemas, transforms can generate erroneous output due to incorrect mapping, or one of the functoids can fault against a particular input. Transforms also support testing with sample data in Visual Studio. Maps can be tested by right-clicking on the map and choosing **Test Map**, as shown in the following screenshot. Any errors during testing are indicated as transform runtime exceptions in the VS **Error List** window. If there are no errors, the output from the transform is indicated in the **Output** tab. If there are errors after the map is deployed during runtime, error tracking records with `xmlTransform` can be seen in the **Tracking** view.

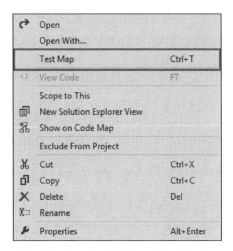

Test Map functionality from Visual Studio

# Troubleshooting bridges

Earlier, we saw how to troubleshoot two stages of bridges, namely the schema validation stage and the transform stage. While using custom code inside a bridge, things can get difficult if the message processing runs into errors. It is recommended that you use `IMessageInspectorContext.Tracer` to log the `System.Diagnostics.TraceEventType` error as part of the custom code. These statements would be surfaced in the WADLogsTable mentioned earlier.

# Troubleshooting agreements

Agreements can either be a transport-level agreement, such as AS2, or a protocol-level agreement, such as X12 or EDIFACT. In B2B scenarios with X12 and EDIFACT agreements, it can happen that the transport status returned to the client sending the message is an HTTP 200 OK, but the message landed in the suspend endpoint. This can happen if there are protocol-level errors. Such errors would be indicated by the acknowledgment message.

Some sample scenarios are shown in the following table:

| Configuration | Scenario | Outcome |
| --- | --- | --- |
| AS2 standalone receive with sync MDN | Incorrect configuration, for example, certificate incorrect | HTTP 400 with error MDN |
| AS2 standalone receive with async MDN | Incorrect configuration, for example, certificate incorrect | HTTP 200 OK for async and MDN with error |
| AS2 standalone send | Incorrect configuration, for example, certificate incorrect | HTTP 500, AS2 Message Sender Activity error |
| X12 or EDIFACT standalone receive | Identities mismatch in the incoming message | HTTP 200 for a client may indicate success, but see the **Tracking** view if the message has ended up in the suspend destination with the error, NACK, to be sent back if configured |
| X12 or EDIFACT standalone send | Schema not found | HTTP 200 for a client may indicate success, but see the **Tracking** view if the message has ended up in the suspend destination with the error, NACK, to be sent back if configured |

# Troubleshooting hybrid connectivity

Finally, we wrap up our discussion by looking at troubleshooting hybrid connectivity. Primarily, this involves looking at the BizTalk Adapter Service, which was introduced in *Chapter 4, Enterprise Application Integration*.

To troubleshoot the hybrid connectivity runtime, add the following snippet to `web.config` in `C:\Program Files\Microsoft BizTalk Adapter Service\ BAServiceRuntime` with administrator access:

```
<system.diagnostics>
  <sources>
    <source name=
      "Microsoft.ApplicationServer.Integration.BAService.Runtime"
      switchValue="All">
<!-- Use Critical, Error, Warning, Verbose, All,
  Information to adjust the log level -->
      <listeners>
        <add name="BAServiceRuntimeTrace" />
      </listeners>
    </source>
  </sources>
  <trace autoflush="true" />
    <sharedListeners>
      <add name="BAServiceRuntimeTrace" type=
        "System.Diagnostics.XmlWriterTraceListener"
          initializeData= "C:\logs\RuntimeTraceFile.xml" />
    </sharedListeners>
</system.diagnostics>
...
<system.serviceModel>
...
  <diagnostics>
    <messageLogging
      logEntireMessage="true"
      logMalformedMessages="true"
      logMessagesAtServiceLevel="true"
      logMessagesAtTransportLevel="true"
      maxMessagesToLog="3000"
      maxSizeOfMessageToLog="2000"/>
  </diagnostics>
</system.serviceModel>
</configuration>
```

The listener configures the traces to be output to an XML file in the user's folder specified in the configuration. Post messaging, we can look at the trace log file to check for errors with the Line of Business access or service configuration issues.

# Summary

In this chapter, we have looked at the ways to collect data to troubleshoot BizTalk Services. This helps in maintaining the health of the service. We also looked at the error scenarios of the key components in BizTalk Services and ways to troubleshoot them.

Troubleshooting is as much an art as it is a science and usually involves a methodical approach to identify and fix a problem. In the next and final chapter, we will look at migration and also capabilities that could be added in the Integration platform.

# 8

# Moving to BizTalk Services

In this final chapter, we will discuss how to move to BizTalk Services. All through this book, we've looked at the new features of BizTalk Services and what they enable, but the likelihood is that you'll be wanting to move existing solutions to BizTalk Services. As you're reading this book, we'll make a further assumption that you want to know more about moving BizTalk Server solutions on-premises to BizTalk Services.

In this chapter, we'll look at the following topics:

- What's available to help move BizTalk Server solutions to BizTalk Services
- How to deal with differences between the products
- Future plans for BizTalk Services

By the end of this chapter, you should have a good understanding of how to tackle moving the existing BizTalk Server solutions to BizTalk Services and what is planned in the evolution of BizTalk Services to make this even easier.

# Moving from BizTalk Server

BizTalk Server consists of a number of architectural components, only some of which have parity in BizTalk Services. These are listed in the following table, which shows the comparison of BizTalk Server and BizTalk Services:

| BizTalk Server | BizTalk Services |
| --- | --- |
| Map | Transform |
| Pipeline | Bridge |
| Business Rules Engine | No equivalent / custom coding needed |
| Business Activity Monitoring | No equivalent / custom coding needed |
| Orchestration | Workflow outside of service |
| Adapters | Bridge sources and destinations |
| Schema | Schema |
| Tracking | Tracking |
| Trading Partner Management | Trading Partner Management |

As you can see from the preceding table, there are a number of functional differences between BizTalk Services v1.0 and BizTalk Server. This is to be expected, as BizTalk Server is an established, mature product that was first shipped in 2000; it has received numerous updates over time. BizTalk Services, on the other hand, was GA'd (released for general availability) on November 21, 2013. However, all is certainly not lost, as there are several ways to mitigate the effort of moving from one to the other. In the following sections, we'll look at the different types of artifacts in a BizTalk Server solution and how to migrate to BizTalk Services.

# Maps

*Chapter 2, Messages and Transforms*, covered mapping in detail and mentioned a tool we'll look at more closely here. First though, let's answer the question of whether you even need a tool.

BizTalk Server provides the ability to run custom **Extensible Stylesheet Language Transformations** (**XSLT**)—just provide the path to an XSLT template file and the map contents are ignored. Maps written in this way do not need the conversion tool as you can simply take the XSLT and configure it in the same way in a BizTalk Services transform.

The functional equivalent of maps in BizTalk are transforms in BizTalk Services, as you have already seen. The implementation of these technologies is, however, very different, and it is not possible to execute BizTalk Server maps in BizTalk Services.

Microsoft has released a conversion tool that takes a BizTalk map, a `.btm` file, as the input, and outputs a BizTalk Services transform file, `.trfm`. Let's see how this works.

In *Chapter 2*, *Messages and Transforms*, we looked at a BizTalk Services transform and the source and target schemas for it. Here, we'll revisit these schemas and look at the equivalent original BizTalk Server map. This is shown in the following screenshot:

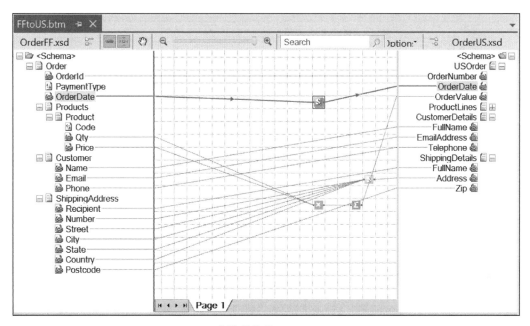

A BizTalk Server map

The conversion tool is part of the BizTalk Services SDK available at `http://www.microsoft.com/en-us/download/details.aspx?id=39087`. Select the `Tools.zip` download and unzip it on your local machine. This is a command-line-driven tool, so here, I'll open a command window as shown in the following screenshot. The executable takes only two parameters: the path to the BizTalk `.btm` file and the output `.trfm` file (which is actually optional).

The BizTalk map conversion tool

The resultant BizTalk Services transform is shown in the following screenshot. Running this yields exactly the same output as the transform we created in *Chapter 2, Messages and Transforms*.

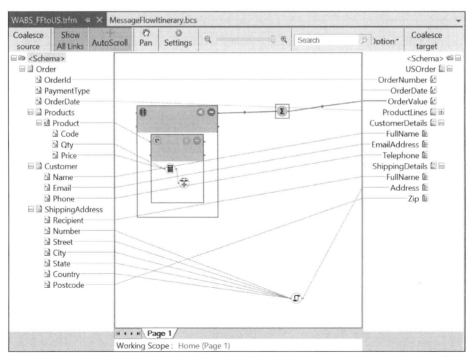

A converted map

As you've just seen, the conversion tool is a useful way to reuse existing maps in BizTalk Services. It is often the case that a large investment has been made in BizTalk maps, and this tool allows them to be converted with minimal effort in many cases. However, the tool is not without a few limitations that will require some rework to the resultant transforms.

One fix that may be required to the generated transform if it won't load correctly is to change the ID values in the file as it can sometimes emit duplicates. When you attempt to open the map in Visual Studio, you may see the error shown in the following screenshot:

The mapping error message

This should be a simple case of finding the ID that is flagged as a duplicate when you try and open the transform and substituting its value with a new, unique one. If the ID refers to a script, there will be at least two places where the duplicate ID will need to be changed. An example is shown in the following code where the contents of the `Value` tag contain a duplicate ID. Changing this to a unique value, for example, `14`, will fix this in the sample accompanying this chapter.

```
<a1:KeyValueOfstringanyType>
    <a1:Key>Id</a1:Key>
    <a1:Value i:type="xs:int">13</a1:Value>
</a1:KeyValueOfstringanyType>
```

Generally, if the tool cannot convert the map, it will convert as much as it can and substitute functoids that cannot be converted with an arithmetic expression functoid in the transform. This will be empty and therefore won't compile to indicate that you need to review it. If a functoid conversion is not possible, the equivalent functoid in the transform will have no inputs to it, again to indicate that something has not been converted. A full list of current limitations is provided in the `ReadMe.txt` file in the same location as the tool itself.

The tool emits a logfile indicating the steps it took to convert the map. The `Log.txt` file will be written to the folder in which the tool was run in the command line.

# Pipelines

Pipelines in BizTalk Server are used to process data between adapters and the MessageBox. This requirement is now provided by bridges in BizTalk Services. We've covered bridges in detail elsewhere in this book (*Chapter 3, Bridges* and *Chapter 4, Enterprise Application Integration*). You should find that usage of bridges and the ability to deploy custom code in them provides most of the functionality you would encounter with the pipelines and custom pipeline components. The BizTalk Services standard bridge stages match quite closely to those of BizTalk Server, as illustrated in the following table:

| BizTalk Server stage | BizTalk Services stage |
| --- | --- |
| Receive: Decode | Enrich (1) |
| Receive: Disassemble | Message Type / Enrich (2) |
| Receive: Validate | Validate |
| Receive: ResolveParty | N/A |
| Send: Pre-assemble | Enrich (1) |
| Send: Assemble | Enrich (2) |
| Send: Encode | Send Reply |
| Receive and Send: Port maps | Transform stage |

If the stages in the preceding table seem a little arbitrary, it's because they are. Just as in BizTalk Server it was perfectly possible to do everything in a single stage (depending on which stage it is of course), so is the case with BizTalk Services. While BizTalk Services doesn't have the concept of pipeline components, it provides enough stage placeholders for your own code and, of course, provides many of the things out of the box for which you'd have traditionally written a custom pipeline component anyway, such as property promotion. Even though BizTalk Server provides four receive pipeline stages, and in fact, pipeline stages are actually configurable (you can define your own), no one really has ever bothered with that and nearly all solutions just use the decode and validate stages (if that). The send pipeline is even less important, but again, BizTalk Services provides a similar mirror image of stages in which you are able to perform work with the messages flowing through should you need to.

All this being said, it should be clear that if you encounter a solution that has custom pipeline components, you have some work to do, that is, to try and convert to BizTalk Services.

# Schema

Schema support is largely the same in BizTalk Services, and generally, you should not encounter too many issues moving your schemas between the two. There are some notable exceptions such as the ability to pass multiple schemas into BizTalk Server maps, which is currently not possible with BizTalk Services, but generally, schemas can be reused in a straightforward manner.

# Adapters

BizTalk Services shares the same concept of adapters, albeit with a much smaller set. Additionally, BizTalk Services provides two approaches to integration: sources and destinations for a bridge as well as the BizTalk Adapter Service, which uses the Service Bus relay to pass messages to an Internet Information Services (IIS)-hosted Line of Business (LOB) adapter (actually, the same LOB adapters BizTalk ships with). Therefore, if the adapters your BizTalk Server solution needs are represented in the BizTalk Services set, the conversion is straightforward. However, there are hundreds of adapters available for BizTalk Server and a dozen or so for BizTalk Services, so clearly, there are some gaps. Some just won't make sense for a cloud-hosted platform (for example, the much-loved File adapter), but for others, this could present a problem.

Microsoft recognizes this potential issue and will, of course, seek to introduce new sources and destinations over time based on customer feedback. Microsoft is also planning to open up the architecture to provide an adapter SDK to enable you (or third parties) to build your own, thus offering another solution. Therefore, over time, this problem is likely to diminish.

# Trading Partner Management (TPM)

BizTalk Server uses TPM to define and manage EDI trading partners. Organizations that use BizTalk's EDI capabilities will likely have hundreds or even thousands of trading partners set up in BizTalk and will want a way to migrate these partners to BizTalk Services if they wish to adopt it.

BizTalk Services takes the same approach to partner management as BizTalk, and changes made in BizTalk Server 2010 to the TPM model and schema were adopted in BizTalk Services. What this means is that the two are actually very similar and migration can be accomplished in a couple of ways.

For existing BizTalk Server users, the TPM Data Migration Tool is provided by Microsoft. This is included in the same `Tools.zip` download as the map conversion tool discussed earlier, and it is capable of migrating the TPM data from either BizTalk Server 2010 or 2013.

The BizTalk Server management console is shown in the following screenshot. Here, you can see two parties and an agreement that we want to migrate to BizTalk Services.

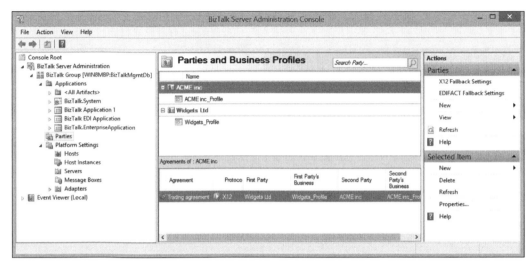

The BizTalk Server management console

To launch it, double-click on `TPMMigration.exe` and the application will appear as shown in the following screenshot:

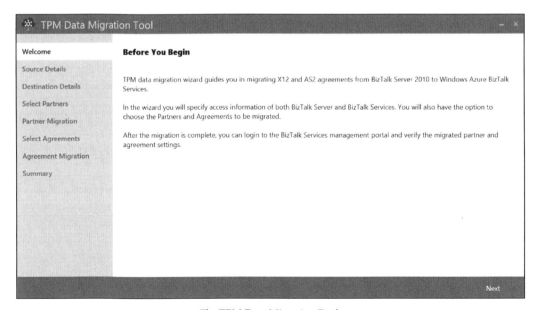

The TPM Data Migration Tool

The tool takes the SQL Server machine name for where the BizTalk Server management database resides and the ACS connection details for BizTalk Services as the input. It then displays the available partners to migrate as shown in the following screenshot:

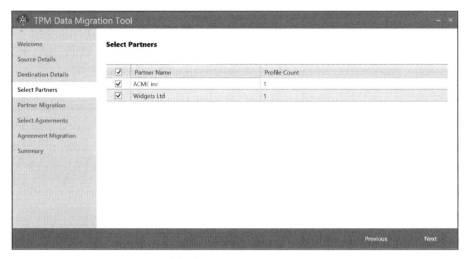

Selecting partners to migrate

Here, I've picked the two partners from the BizTalk Server management console. Clicking on **Next** then displays the agreements. I only have one, shown in the following screenshot, which is between the two parties, so selecting this and clicking on **Next** will start the process of moving the partners and agreements to BizTalk Services:

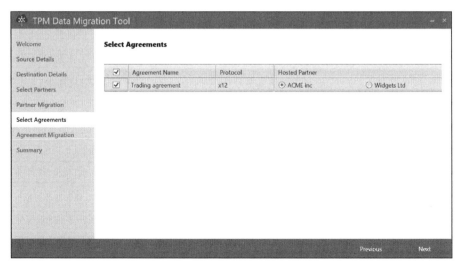

Selecting agreements

Once the partners and agreements have been created in BizTalk Services, a **Summary** page details the work completed, as shown in the following screenshot:

The migration summary

Now, looking in the BizTalk Services portal, we can see that the partners and agreement have been created, as shown in the following screenshot:

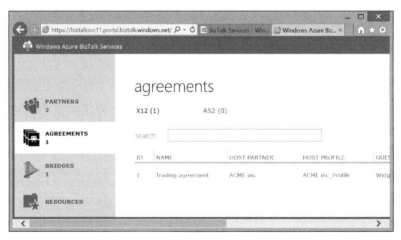

The BizTalk Services portal

The tool should be able to migrate all the parties and agreements you have set up in BizTalk Server. The one limitation the tool has is that it won't move the certificates that you use to secure the conversations between your organization and the trading partners. These need to be migrated manually.

The TPM Data Migration Tool is useful when moving from BizTalk Server 2010/2013 to BizTalk Services. However, another option exists if you need to move from a different product or an earlier version of BizTalk Server. This option is also a useful approach if you want to programmatically create trading partners perhaps, from a custom application or via integration with another product. This approach uses the TPM API. In fact, the migration tool leverages this API as well to do its jobs.

Previously, the API that TPM used was not documented, and as such, customers were not supported if they wished to create trading partners in BizTalk programmatically. This is no longer the case. Microsoft has now published the API on MSDN, thus allowing customers to leverage it in a supported way.

The BizTalk Services TPM API is documented at the following location:

```
http://msdn.microsoft.com/en-us/library/windowsazure/dn232369.aspx
```

In order to call the TPM API, an OAuth WRAP token is necessary for authentication. This token is simply a string consisting of the following pieces of information:

- User name: owner
- Password: Issuer Key from ACS
- BizTalk Services endpoint name: https://<yourservice>.biztalk. windows.net/

The process of calling the API is made up of two steps. First, POST a WRAP request and receive a WRAP token, which is then passed in on subsequent requests. The API is REST-based just like the other APIs we looked at in *Chapter 6, API*. It is harder (although not impossible) to invoke them in Fiddler (or the browser) because of the OAuth requirement (as opposed to mutual certificate authentication). Therefore, let's look at the code necessary to retrieve a list of partners as an example.

The following code will call Azure with the WRAP request and obtain a token:

```
string nameSpace = "<your WABS  namespace>"; // WABS namespace
string defaultIssuer = "owner"; // WABS issuer - usually "owner"
string defaultKey = "<your WABS key>"; // WABS issuer key
string serviceName = "gettingstartedwabs";
string address = string.Format((IFormatProvider)CultureInfo.
InvariantCulture, "https://{0}.{1}/{2}/",
nameSpace, "accesscontrol.windows.net", "WRAPv0.9");
string payload = string.Format((IFormatProvider) CultureInfo.
InvariantCulture,
```

```
"wrap_name={0}&wrap_password={1}&wrap_scope={2}", defaultIssuer,
                Uri.EscapeDataString(defaultKey),
                Uri.EscapeDataString("http://" + serviceName +
".biztalk.windows.net/default/$PartnerManagement/Partners"));
            HttpContent content = new StringContent(payload);
            content.Headers.ContentType.MediaType = "application/x-
www-form-urlencoded";
using (var client = new HttpClient())
{
    // get WRAP token
    var response = await client.PostAsync(address, content);
    response.EnsureSuccessStatusCode();
    string token = await response.Content.ReadAsStringAsync();
    token = Uri.UnescapeDataString(token.Split('&')[0]);
}
```

This is quite straightforward. There are four pieces of information required. To obtain the ACS details for your BizTalk Services instance, go to the Azure portal, click on **BizTalk Services** in the left margin, select your instance, and click on **Connection Information**. You will find the namespace, issuer, and key here to substitute in the preceding code. The service name is the name you gave the BizTalk Services instance when you created it and will be the title displayed on the Azure portal dashboard.

This information is concatenated and sent to ACS. It validates and returns an authentication token—a string that can be used on subsequent calls.

The following piece of code (which should be placed just inside the preceding closing curly brace of the using statement) will pass in the token with the request and receive a list of partners in the specified BizTalk Services instance:

```
// get partner list
client.DefaultRequestHeaders.Add("x-ms-version", "1.0");
client.DefaultRequestHeaders.Authorization = new AuthenticationHeaderV
alue("WRAP", "access_token=\"" + token.Substring(18) + "\"");
response = await client.GetAsync("https://" + serviceName + ".biztalk.
windows.net/default/$PartnerManagement/Partners");
// write out partner list
Console.WriteLine("Partners:");
System.Xml.XmlDocument doc = new XmlDocument();
doc.LoadXml(await response.Content.ReadAsStringAsync());
foreach (XmlElement node in doc.SelectNodes("//*[local-
name()='feed']//*[local-name()='entry']//*[local-
name()='content']//*[local-name()='properties']//*[local-
name()='Name']"))
{
    Console.WriteLine(node.InnerText);
}
Console.ReadLine(); // wait
```

The preceding code performs an HTTP GET request on your BizTalk Services instance endpoint, appending the operation ($PartnerManagement/Partners) and passing the token. The response is an XML document containing all the partners set up in the BizTalk Services instance. To try this out, create a console application in Visual Studio and paste the code in the Main method, replacing the values marked in the code with your own service details. The resulting output from the partner import performed earlier is shown in the following screenshot:

List of partners

There is much more that can be done with the API, such as creating partners and updating or deleting them. However, the approach is always the same, so feel free to explore on your own and see what you can do!

# EDIFACT support

BizTalk Services first shipped with support for X12 and AS/2. As of the February 2014 service update, EDIFACT support is now also provided and will be particularly welcome for European customers.

# Business Rules Engine (BRE)

Now we come to a few areas that are more problematic. BizTalk Server has provided a Business Rules Engine and editor since 2004, and as such, it is used in many BizTalk Server solutions. There is no equivalent in BizTalk Services currently.

Microsoft is planning to provide a rules engine as part of BizTalk Services at some point, but there is no timeline for it yet. The intention is to provide parity with BizTalk Server and improved tooling, and both of these developments will make moving from Server to Services easier when introduced.

In the meantime, one option is to convert BizTalk BRE rules to code. There are a number of solutions available that are able to convert BizTalk rules to Windows Workflow rules, and Windows Workflow rules are defined in code. Windows Workflow is also a part of the .NET framework, so there is no license cost in using it. Therefore, it is possible to run the code somewhere in the BizTalk Services solution, for example, in a bridge or transform. Of course, this is trivializing the problem slightly as BRE rules can access databases and other resources, so it may involve significantly more work than simple conversion. However, it is an option depending on what the rules are doing.

# Orchestration

Probably the biggest challenge in moving BizTalk Server solutions to BizTalk Services is orchestration. Currently, there is no silver-bullet, automatic, or zero-effort way to convert or migrate orchestrations to BizTalk Services. There are some options though.

Microsoft plans to introduce workflow to BizTalk Services, and this will certainly help fill the gap. It means that orchestrations can be recoded to workflows and keep a similar architecture.

First, remember that a bridge is actually a workflow. This means that bridges already offer some capabilities that orchestrations may have been used for earlier, such as message enrichment and routing (which is a large percentage), and as such, it may already be possible to migrate orchestration-based solutions.

In the intervening period though, an alternative solution may have to be found. One solution is to use a workflow hosted in Azure, for example, in a worker role cloud service. BizTalk Services could invoke the cloud service by passing a message or data, and the service would run the workflow and return the results. This does change the architecture somewhat though as typically, the orchestration is in control—it may wait for a set interval or for specific responses from other systems and typically, an orchestration is used as the driver for a business process. It is worth remembering that bridges can be chained, so this style of process definition can be mimicked with BizTalk Services, where messages are processed, decisions are taken on routing (to more bridges), and so on. However, such a solution is likely to become complex and is best avoided.

It is perhaps unfortunate then that an overuse of orchestration in BizTalk Server has always been prevalent. Orchestration was seen as the "aha" moment in early BizTalk when acronyms such as Business Process Management (BPM) were in vogue. It's unfortunate because orchestration has often been used when it is not necessary, and a simpler solution could have been created without it. While education over time has helped somewhat, there is still a large body of complex orchestration-centric BizTalk Server applications. If we face this scenario, migration to BizTalk Services today will be challenging.

# When not to move

Before closing, its worth pointing out that BizTalk Services is not meant to replace BizTalk Server. While there are certainly many similarities in the capabilities offered (and more on the way), there are different reasons for using each. Here are some of the reasons you should continue using BizTalk Server on-premises:

- All of your connectivity points (applications, services, and so on) are on-premises

- A large investment in BizTalk Server-specific solutions — as this would be likely to require a complete rewrite, therefore, outweighing some of the benefits

- Usage of capabilities that are not in BizTalk Services or will not fit the cloud model, for example, File and MQ Series adapters

- Cloud is not the right solution because for example, security and/or regulatory restrictions, data classifications, or local laws may preclude sending such data over the public Internet or prevent storage of data off-premises

# The future

Microsoft has committed to continuing investments in BizTalk Server and a strong roadmap for BizTalk Services. The following are the key announced developments coming for both products:

- BizTalk Server will ship a major release every alternate year
- A platform update release of BizTalk Server will ship every other year starting with BizTalk Server 2013 R2 this year
- The following additions are planned for BizTalk Services with a target of a quarterly release cadence:
    - Workflow integration
    - Rules engine integration
    - Business Activity Monitoring
    - Adapter extensibility/SDK
    - Custom code improvements in bridges
    - Integration with Windows Azure Active Directory (WAAD)
    - Business Process Modelling Notation (BPMN) support
    - Windows Azure store for third-party components
    - Scheduled backups

# Summary

In this chapter, we've looked at strategies and approaches for moving from BizTalk Server to BizTalk Services and some of the features that will be added to BizTalk Services that will make this easier. We tried to cover all the main building blocks of the BizTalk Server architecture and their equivalents (or alternatives) in BizTalk Services. As you've seen, while there is a significant overlap of functionality, it will take time for the BizTalk Services feature set to mature to the same level. The future for BizTalk Services is bright, and a fundamental tenet of Microsoft's cloud-first vision is to deliver frequent updates and enhancements in a way that simply isn't possible with shrink-wrapped on-premises products. Therefore, one can expect Microsoft to become even more customer focused and be able to action feedback and feature requests quicker than previously possible, all of which will enable Microsoft to help you move your existing solutions to the cloud.

# Index

## Thank you for buying
# Getting Started with BizTalk Services

## About Packt Publishing

Packt, pronounced 'packed', published its first book "Mastering phpMyAdmin for Effective MySQL Management" in April 2004 and subsequently continued to specialize in publishing highly focused books on specific technologies and solutions.

Our books and publications share the experiences of your fellow IT professionals in adapting and customizing today's systems, applications, and frameworks. Our solution based books give you the knowledge and power to customize the software and technologies you're using to get the job done. Packt books are more specific and less general than the IT books you have seen in the past. Our unique business model allows us to bring you more focused information, giving you more of what you need to know, and less of what you don't.

Packt is a modern, yet unique publishing company, which focuses on producing quality, cutting-edge books for communities of developers, administrators, and newbies alike. For more information, please visit our website: www.packtpub.com.

## About Packt Enterprise

In 2010, Packt launched two new brands, Packt Enterprise and Packt Open Source, in order to continue its focus on specialization. This book is part of the Packt Enterprise brand, home to books published on enterprise software – software created by major vendors, including (but not limited to) IBM, Microsoft and Oracle, often for use in other corporations. Its titles will offer information relevant to a range of users of this software, including administrators, developers, architects, and end users.

## Writing for Packt

We welcome all inquiries from people who are interested in authoring. Book proposals should be sent to author@packtpub.com. If your book idea is still at an early stage and you would like to discuss it first before writing a formal book proposal, contact us; one of our commissioning editors will get in touch with you.

We're not just looking for published authors; if you have strong technical skills but no writing experience, our experienced editors can help you develop a writing career, or simply get some additional reward for your expertise.

## Microsoft BizTalk ESB Toolkit 2.1

ISBN: 978-1-84968-864-2          Paperback: 130 pages

Discover innovative ways to solve your mission-critical integration problems with the ESB Toolkit

**Microsoft BizTalk ESB Toolkit 2.1**

Discover innovative ways to solve your mission-critical integration problems with the ESB Toolkit

Andrés Del Rio Benito   Howard S. Edidin

1. A comprehensive guide to implementing quality integration solutions.

2. Instructs you about the best practices for the ESB and also advises you on what not to do with this tool.

3. A sneak view of what's new in the ESB Toolkit 2.2.

## (MCTS) Microsoft BizTalk Server (70-595) Certification and Assessment Guide *Second Edition*

ISBN: 978-1-78217-210-9          Paperback: 570 pages

Including Microsoft Partner Network Technical Competency Assessment for Application Integration (BizTalk Server 2013) and Windows Azure BizTalk Services coverage

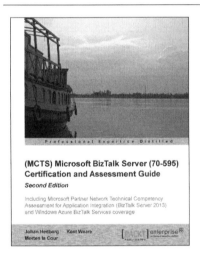

**(MCTS) Microsoft BizTalk Server (70-595) Certification and Assessment Guide**
**Second Edition**

Including Microsoft Partner Network Technical Competency Assessment for Application Integration (BizTalk Server 2013) and Windows Azure BizTalk Services coverage

Johan Hedberg   Kent Weare
Morten la Cour

1. Features a comprehensive set of test questions and answers that will prepare you for the actual tests.

2. The layout and content of the book matches the structure of the exam closely, which maximizes your study time and helps you focus on learning areas where you need improvement.

Please check **www.PacktPub.com** for information on our titles

Made in the USA
San Bernardino, CA
28 March 2014